CIOFORUM

I0482367

# RFP Datacenter Hosting

<date>

Filled in Case Template for Best Value Procurement

| Project details | | |
|---|---|---|
| **ISBN-13:** | 978-1534602786 | |
| **ISBN-10** | 153460278X | |
| **Publish date:** | June, 2016, first print | |
| **Project manager/authors:** | Robert Zondervan, MSc, RZondervan@cioforum.nl | |
| **Publisher and © copyright** | CIOforum.nl | |

## TABLE OF CONTENTS

# 1 DATACENTER MIGRATION RFP

Customer requests to move their self-managed datacenters to a vendor with capability to deliver Infrastructure-as-a-Service.

## 1.1 Management Summary

This Request for Proposal has the objective to start with a *"Lift & Shift"* project of Customer's twin datacenters in the Summer of 20xx, with a realization deadline of &lt;date&gt;, and subsequent move to managed services, including cloud, when opportunities arise. The final decisions as to whether or not to move to managed services are for Customer.

The Customer existing team will still be responsible for datacenter services after the completion of Lift & Shift, except for housing services, such as cooling, power supply and shared connectivity services.

The managed services will be handled by vendor based on a future Customer decision per Lot depending on the recommendation of the vendor and for example taking into account amortization of existing hardware or buy-out of existing hardware. The amortization period used by Customer is 4 years and the purchase date of hardware is given in the As-Is overviews.

### 1.1.1 Part 1 – Lift & Shift

Customer requests a migration plan, including one time migration costs and monthly recurring costs for moving both their datacenters to vendor's datacenters offering the same or better high availability and continuity metrics than the current mode of operation described in 5 lots in chapter 2.

### 1.1.2 Part 2 – Migration to Managed Services when opportunities arise

Customer expects to receive a gradual migration plan, including one time migration costs per lot and monthly recurring costs for each lot, e.g. based on reaching 100% amortization (4 years after investment), or handover for depreciated residual value:

- Datacenter Network services, e.g. guarding, monitoring and reporting vulnerabilities, activity and access
- SAP basis and infra services, including storage (out-of-scope are build & functional services)
- Server virtualization, including storage
- Citrix desktops (which are not attached to storage)
- Mail services

## 1.2 Identified Risks

The identified risks, our biggest worries based on our preliminary market exploration (RFI/RFQ), and influencing the setup of this RFP, are:
- Unexpected migration costs for both:
  - Stage 1: Lift & Shift stage
  - Stage 2: From self-managed technical domain(s) to managed service(s)
- Unexpected future increased costs for yearly projects within managed services and workload for retained team
- Sales promotion not in line with delivery (vendor lock-in)

Risk Mitigations:

- Customer delivers an exhaustive As-Is Inventory description and requests fixed price offers for migrations and Customer calculates assumed prices for yearly projects and growth.
- Defining roles & responsibilities together with vendor, during the RFP stage.
- Final award stage with preferred supplier(s): Create project plan, contract, fixed migration pricing for both L&S and managed services, and SLA documents ready for sign-off & delivery.
- Interviews/demos with key engineers (do we get what is promised?).
- Enablement of Managed Services by potentially multiple third parties or cloud.

### 1.3 Best Value Procurement – Weigh factors

A. Pricing (40%)
B. Quality (30%)
C. Offered HR track (30%)

Intermediate short listing based on vendor's capability assessment on partial assessments of pricing, quality, and HR capabilities after 5 Vendor Activity stages: pre-V1, V1, V2, V3 and V4.

**Proof of capability** documents of the Pre V1 stage:

- Deliver company information, shareholders, structure of mother corporation, legal forms, joint ventures, subsidiaries, partnerships or other relevant relations
- Deliver complete Managed Services SLA example set, incl. gov., custom procedures
- Deliver standard terms & conditions
- Deliver standard maintenance windows of managed services (planned down time calendar)
- Deliver HR track capability: CLA 32bis experience and references (template D in V3)
- Demonstrated Capabilities Form (References) (template B to be used in V1)

Final awarding:

| | | | | |
|---|---|---|---|---|
| A. | Pricing (template A) | | | 40% |
| | A1. | Pricing - Lift & Shift | | |
| | A2. | Pricing - Managed Services | | |
| | A3. | Pricing - Calculated Project Assumption | | |
| | A4. | Pricing - Calculated Assumptions for growth and hosting | | |
| B. | Quality - Lift & Shift (L&S) | | | 10% |
| | B1. | L&S - Description of housing / colocation services | 10% | |
| | B2. | L&S - Project Organization | 10% | |
| | B3. | L&S - Project Methodology, Tools, Technology | 20% | |
| | B4. | L&S - Demonstrated Capabilities and References (template B, (pre) V1) | 10% | |
| | B5. | L&S - Risk File (template C1) | 20% | |
| | B6. | L&S - Project Plan & Roadmap (V2) | 30% | |
| C. | Quality - Managed Services (MS) | | | 20% |
| | C1. | MS - Organization | 5% | |
| | C2. | MS - SLA, incl. maintenance windows (Pre V1 - V4) | 10% | |
| | C3. | MS - Description of services | 15% | |
| | C4. | MS - Migration Project Organization | 5% | |
| | C5. | MS - Project Methodology, Tools, Technology | 5% | |
| | C6. | MS - Demonstrated Capabilities and References (template B, (pre) V1) | 10% | |
| | C7. | MS - Risk File (template C2) | 5% | |
| | C8. | MS - Capabilities and competences per technical domain | 10% | |
| | C9. | MS - Opportunities Form (template E) | 5% | |
| | C10. | MS - RACI's, policies, procedures (V2) | 10% | |
| | C11. | MS - Project Plan & Roadmap (V3) | 10% | |
| | C12. | MS - Interviews and PoC / demo (between V3-V4) | 10% | |
| D. | HR Track | | | 30% |
| | D1. | HR - CLA 32bis Experience and References (Pre V1 - V4) (template D) | 100% | |

## 1.4 RFP Project Timeline with vendor activity periods V1-V4

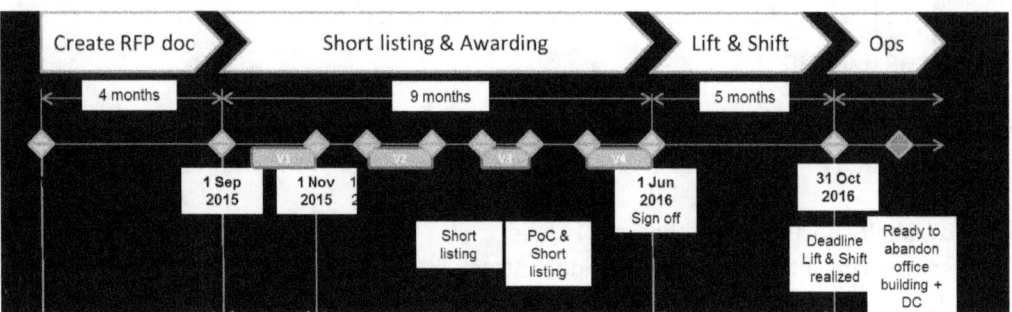

*Image: Project Timeline with vendor activity periods V1-V4*

### 1.4.1    Short listing stage – Pre V1 activities – Proof of Capability

The Proof of Capability is used to select and filter initial vendors for the long list, starting with V1.

**Focus on HR capability**

Except the proven capabilities around datacenter services, there is a strong focus on vendor's HR capabilities and vendor experience with the Belgian Collective Labor Agreement 32bis in order to be able to offer, a selection of future potential obsolete existing internal IT engineers, a new employer.

| ID | Date | | Vendors |
|----|------|--|---------|
| V1.1a | Before 1 Sep 2015 | • Infra due diligence for As-Is description<br>• Create & review RFP documents\*<br>*\*new releases with tracked changes during short listing period*<br>• Define long list | |
| V1.1b | Before end-of-June 2015 | • Inform vendors (e.g. context, initial timeline, vendor activity periods)<br>• Send NDA (if not already signed RFI/RFP version from Q42014) | RFI vendors: Deliver proof of capability before 25 July 2015<br>**Proof of capability**<br>• Deliver company information, shareholders, structure of mother corporation, legal forms, joint ventures, subsidiaries, partnerships or other relevant relations<br>• Deliver complete SLA example set, incl. gov., custom procedures<br>• Deliver standard terms & conditions<br>• Deliver standard maintenance windows, e.g. quarterly change weekend: Sun 6am-12am (planned down time)<br>• Deliver HR track capability: CLA 32bis experience and references<br>• Demonstrated Capabilities Form (References) |
| V1.1c | August 2015 | • New Vendor information sessions | • Sign off NDA (if not already signed RFI/RFP version from Q42014)<br>• Information sessions per new vendor (vendors which were not involved in RFI)<br>• New vendors: Deliver "proof of capability" (see above) before 15 August 2015 |

### 1.4.2    Short listing stage – V1 activities

| ID | Date | | Vendors |
|---|---|---|---|
| V1.2 | 4 Sep 2015 | Delivery and communication of complete RFP :<br>• Introduction and timeline<br>• As-Is description (results of due diligence) :<br>    • Lift & Shift section<br>    • Managed Service section<br>• Answer template & Scoring weight | • Vendor gets informed |
| V1.3 | 15 Sep 2015 | Accept/deny RFP candidate based on delivered capability overview of pre V1 | V1: Vendor activity period 1:<br>Accept candidacy before 15 Sep 2015<br>Answer RFP :<br>• Colocation (housing)<br>• Virtual Private Cloud enablement (Azure, AWS, …)<br>• IaaS<br>• Managed Service (MS) Options :<br>    • SAP, Notes, etc.<br>• Pricing for L&S and MS<br>• Roadmap proposal<br>• Available competences per technical domain<br>• Opportunity Form<br>• What does vendor want to know to be able to answer vendor HR track activities in V3? |
| V1.4 | 1 Nov 2015 | - | Due date delivery of RFP V1.3 answer |

### 1.4.3    Short listing stage – V2 activities

| ID | Date | | Short listed Vendors |
|---|---|---|---|
| V2.1 | 1 Nov 2015 | Score RFP answers V1<br>Short list vendors for V2 | - |
| V2.2 | 1 Dec 2015 | Inform vendors about acceptance/denial RFP candidates based on scoring<br>Deliver RFP feedback & update<br>RFI WAN connectivity: 1)▢DC; 2) New DC to MPLS (▢)<br>(price and delivery time) | V2: Vendor activity period 2: Answer RFP :<br>• Managed Service RACI's<br>    (to be able to define workload for retained team)<br>• Roadmap Lift & Shift, plus pricing |
| V2.3 | 15 Jan 2016 | Score RFP answers V2<br>Selection final vendors for V3 | Due date delivery of RFP V2 answer |
| V2.4 | 15 Feb 2016 | Inform vendors about acceptance/denial RFP candidates based on scoring<br>Deliver detailed HR information to V3 vendors. | |

### 1.4.4    Short listing stage – V3 activities

| ID | Date | | Final Vendors |
|---|---|---|---|
| V3.1 | 15 Feb 2016 | Deliver RFP feedback & update | V3: Vendor activity period 3: Answer Managed Services RFP :<br>• For interview without sales presence: Select 2 managed service engineers, who will actually be assigned to ▢<br>• HR track, e.g.<br>    • Pension<br>    • Monetary Value calculation from old to new situation<br>• Roadmap Managed Services, plus pricing (BAFO) |
| V3.2 | 15 Mar 2016 | Score RFP answers V3<br>Selection final vendors for V4 | Due date delivery of RFP V3 answer<br>Between V3&V4: Interviews and Proof of Concept (PoC) / demo (VM delivery, including Amazon/Azure) |
| V3.3 | 15 Apr 2016 | Inform vendors about acceptance/denial RFP candidates based on scoring | |

### 1.4.5   Short listing stage – V4 activities

| ID | Date | | Final Vendor(s) |
|---|---|---|---|
| V4.1 | 15 Apr 2016 | Deliver RFP feedback & update | V4: Vendor activity period 4: Final contract & SLA creation :<br>• Final roadmap Lift & Shift<br>• Proposal roadmap from L&S to MS<br>• Pricing negotiation for both L&S, MS |
| V4.2 | 1 Jun 2016 | | Due date deliverables of RFP V4 |
| V4.3 | 1 Jun 2016 | Sign off<br>Inform vendors | Sign off |
| V4.4 | 1 Jun 2016 | Start implementation Lift & Shift | Start implementation Lift & Shift |
| V5.1 | 31 Oct 2016 | Due date  Lift & Shift realization<br>Continuously: Decide on managed service opportunities | Due date  Lift & Shift realization<br>Gradually move to managed services, when opportunities arise |
| V5.2 | 31 Dec 2016 | Ready to abandon office building + Datacenter (DC)<br>Continuously: Decide on managed service opportunities | Gradually move to managed services, when opportunities arise |

## 1.5 Required document delivery and structure

### 1.5.1   Pre Delivery by vendors to enable initial short listing

Deliver the Pre V1 "proof of capability" documents, mentioned in paragraph 1.4.1, before entering V1.3: the vendor activity period for answering the whole RFP (V1 components only).
- RFI vendors: Deliver "proof of capability" before &lt;date&gt;
- New vendors: Deliver "proof of capability" before &lt;date&gt;

### 1.5.2   Delivery via answering templates

Answering structure including ABC sub numbering is required and defined in this document for pricing, quality and HR track. The Vendor may reuse this document (only for answering this RFP).

After Vendor Activity Period 1, vendor is requested to deliver the same answer document twice: one of the previous period (in docx format) with tracked changes, the other without (in pdf format).

## 1.6 Potential Managed Service Domains

The following diagram shows a high level summary of the As-Is situation (July 2015).

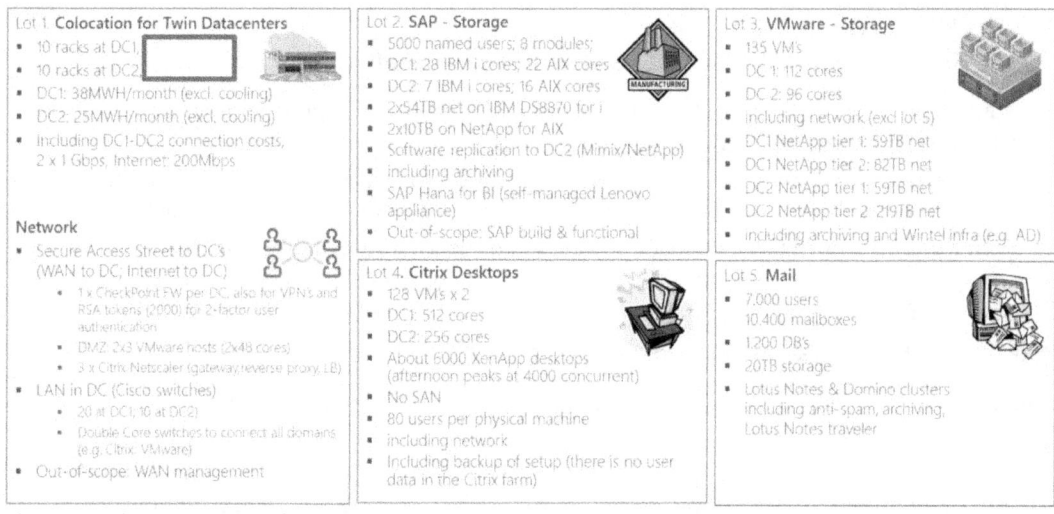

The central Customer end-user Service Desk is out-of-scope and will act as point of contact to vendor's support.

### 1.7 Communication Plan

All mail with CC to:

- RZondervan@cioforum.nl :
  Project Manager datacenter governance & architecture
- Bert@Customer
  Infrastructure Manager
- Werner@Customer
  Vendor & License Manager

Customer will share (anonymized) questions & answers with all RFP vendors

## 2  DESCRIPTION OF AS-IS SITUATION

1.  Specific SLAs can be found in the **CUS service catalogue.pdf**

2.  Support hours and shifts can be found in the CUS Service catalogue in combination with the **Major incident procedure.docx**. In principle the systems need to be up and running 24/7 and we have an on duty system with system engineers in case of major incidents. All P2 incidents will be treated between 07:00 and 23:00 CET and all P1 incidents will be covered with the 24/7 availability of our service desk and the on duty system engineers.

### 2.1.1  IT FTE's

How many IT FTE's (total and per domain)

| Domain | As-is Total | As-is Internal |
|---|---|---|
| **DC Facilities** | | |
| **SAP Infra and Basis** | 3 + Ad hoc | 2 (shared resources) |
| **Wintel (VMware, Storage, Citrix)** | 5 + Ad hoc | 4 |
| **Network Security** | 2 + ad hoc | 2 (shared resources) |
| **Email** | 2 + ad hoc | 1 |
| **Projects Manager** | 1 | 1 |

All system engineers do support and project activities. For 2015, we will have approx. 900 man days for projects done by internal resources. Specific external resources for projects are not included.

### 2.1.2  Information Library

| Context | File name | Content |
|---|---|---|
| **Service Catalogue** | Service catalogue.pdf | |
| **Major incident procedure** | Major incident procedure.docx | Description of P1 and P2 incident handling |

### 2.2 LOT 1 – DATACENTER

### 2.2.1  Datacenter room DC1 – Cus

The production datacenter is located in the Etercenter in Location 1 (CUS) on the first floor. There is a need to use a crane to lift the datacenter equipment to move the equipment from the 1st floor of the Office building.

The datacenter is physically secured by a locked door that can only be opened with a badge of which logging is done. Only a limited number of people (System engineers, Customer Internal helpdesk and Customer Infrastructure management) is authorized to enter the datacenter. Visitors need to sign the log book to enter the datacenter and can only enter when accompanied by authorized personnel.

The datacenter is equipped with a redundant (2) UPS (Galaxy 5000) and a single diesel generator. The UPS can give up to 20 minutes power and the Generator up to 48 hours without refilling the tank. The generator becomes active after 5-10 seconds while the UPS guarantees power to the IT equipment. The airco is only connected to the generator and not to the UPS.

Currently we have 15 racks in place but 5 racks can be consolidated into the existing others so we will reduce the number of racks to 10.

### 2.2.1.1 Information Library

| Context | File name | Content |
|---|---|---|
| **Floor Plan Datacenter** | DC CUS.xls | |
| **Rack Layout Datacenter** | Rack Layout CUS.xlsx | Overview of hardware per rack |
| **Power Schema** | Power Supply for DC1.pptx | Setup Servers – UPS – Generator |
| **Power Usage** | Power Usage CUSTOMER.xlsx | Power usage of Datacenter CUS in 2015 |

### 2.2.2    Datacenter room DC2 – Colocation

The second datacenter is located in Colocation where floor space is rented.
This datacenter hosts the backup and DR systems as well as certain development and test systems for the SAP environment.

The datacenter is physically secured and only a limited number of people (Customer Internal helpdesk and Customer Infrastructure management) is authorized to request access to the datacenter for others.
Visitors can only enter when accompanied by authorized personnel.

Currently we have 10 racks in place of which one is empty. Reorganization is ongoing so we could reduce to 9.

### 2.2.2.1 Information Library

| Context | File name | Content |
|---|---|---|
| **Floor Plan Datacenter** | DC COLO.jpg and DC COLO1.jpg | |
| **Rack Layout Datacenter** | Rack Layout COLO.xlsx | Overview of hardware per rack |
| **Colocation Datacenter** | Colocation Data Center description-V1.11.pdf | |

| Power Usage | Power Usage COLO.jpg | Power usage of Datacenter COLO in 2015 |
| --- | --- | --- |

### 2.2.3 Datacenter Connectivity

Both datacenters are interconnected with 2x 1 Gbps connections from different providers (…).

Both Datacenters are connected to our MPLS provider Provider with a 155 Mbps connection.

Both Datacenters are connected to internet with a 200 Mbps connection (currently 100 Mbps but upgrade will be done soon).

### 2.2.4 Datacenter LAN and Firewall

We have 2 core switches in CUS in VSS (NPSBE03000) - WS-C4500X-16 which are the heart of the infrastructure. The User distribution switches, the MPLS routers, the Internet firewall and the Datacenter firewall are linked to them. The 2 core switches are placed in the 2 different Datacenters in CUS to guarantee connectivity for Eternit (company across the street of the DC) to the datacenter in Colocation in case of failure.

The Datacenter firewall are Checkpoint firewall's build as a Cluster with 2 members in CUS and a 3rth Member in the DRP site. The firewalls are Active/Standby. They are connected with 10 Gbps to each other.

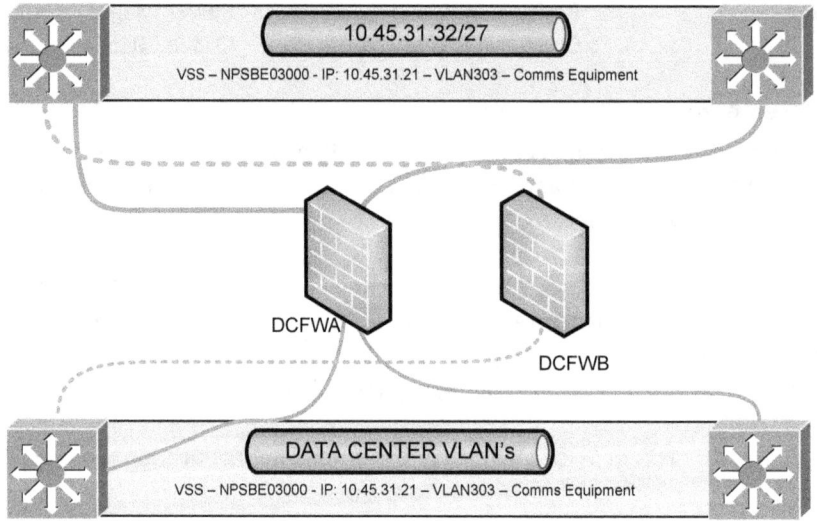

### 2.2.5 Distribution Switches

From the Core equipment we go to 2 Cisco 3750 distribution switch stacks:

- NPSBE03004, 3 Switches
- NPSBE03006, 5 switches

To these stacks we connect all Bladecenter switches and PizzaBox Servers as well as the AS 400 servers. Only exception is the NetApp, which is directly connected to the Core for 10 Gbps connectivity. Currently all devices are connected with at least 2 connections, to 2 different switches, however not yet to the 2 different stack. This is however a change that is in progress. Currently 1 bladecenter is connected this way and a number of servers. (For clarity the 2 different types of connectivity are displayed in blue, the green connections are the as is situation)

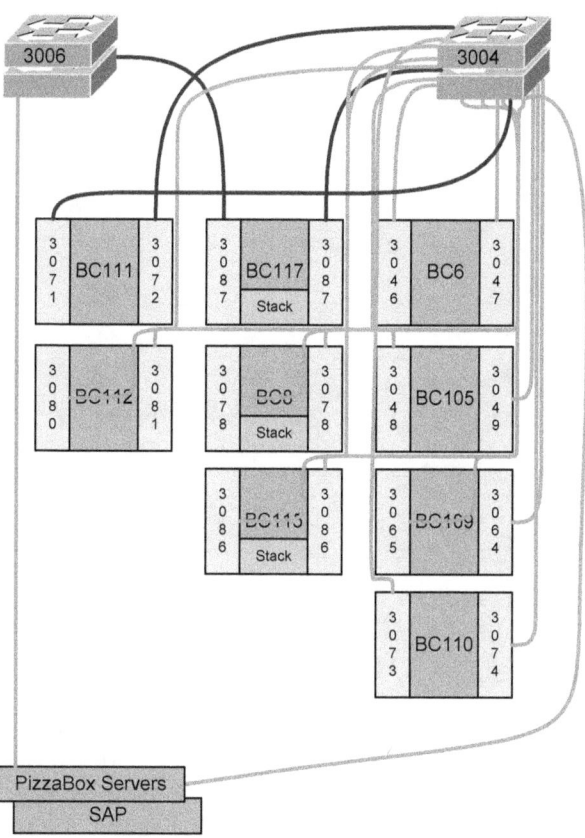

### 2.2.6    Bladecenter Switches Overview

BC117: NPSBE03087 – 10.45.6.187
      (2 x Cisco 3120X)

BC112: NPSBE03080 – 10.45.6.180 / NPSBE03080 – 10.45.6.181
      (2 x Cisco Catalyst 3020)

BC006: NPSBE03046 – 10.45.6.246 / NPSBE03047 – 10.45.6.247
      (2 x Cisco IGESM)

BC008: NPSBE03078 – 10.45.6.178
      (2 x Cisco Switch Module 3110X )

BC110: NPSBE03073 – 10.45.6.173 / NPSBE03074 – 10.45.6.174
      (2 x Cisco Catalyst 3020)

BC111: NPSBE03071 – 10.45.6.171 / NPSBE03072 – 10.45.6.172
  (2 x Cisco Catalyst 3020)
BC109: NPSBE03064 – 10.45.6.164 / NPSBE03065 – 10.45.6.165
  (2 x Cisco Catalyst 3020)
BC105: NPSBE03048 – 10.45.6.248 / NPSBE03049 – 10.45.6.249
  (2 x HP Switch - GbE2c Ethernet Blade Switch)
BC115: NPSBE03086 Stacked – 10.45.6.186
  (2 x Cisco 3120X)

### 2.2.7   DMZ Security

Customer has its own public IP range. This allows us to change internet providers with minimal impact to the end users. This would allow us also to use different Internet Providers on the Main DC and in the DR DC.

Behind the Internet providers Routers we have a cluster of 2 Palo Alto Firewall's (active/passive). 1 member is in the DC in CUS and the other one in is in the DC in Colocation. These firewalls are responsible for the first line of defense. Currently they do antivirus and NAT. After this line, we have Checkpoint Firewall's. The Checkpoint firewall is responsible for IPS and VPN Concentrator for end users and Site2Site VPN's.

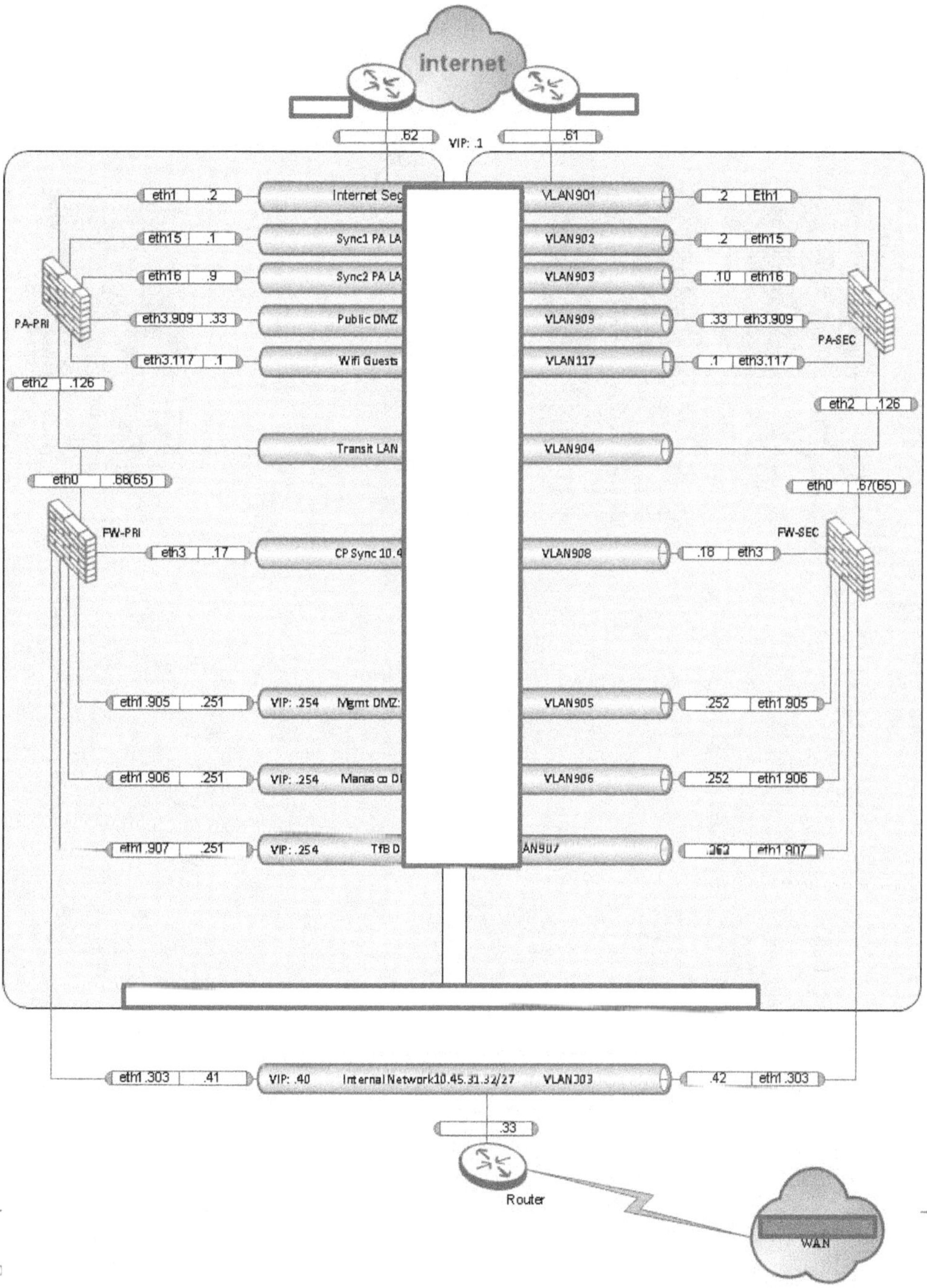

## 2.2.8   DMZ Servers

The VMware cluster for our DMZ consists of 2 hosts in the datacenter CUS and uses storage from NetApp (NetApp storage FAS2552 in CUS and a FAS2240 in Colocation). The Virtual Machines are mirrored to the datacenter in Colocation using NetApp snapshot and mirroring technology.

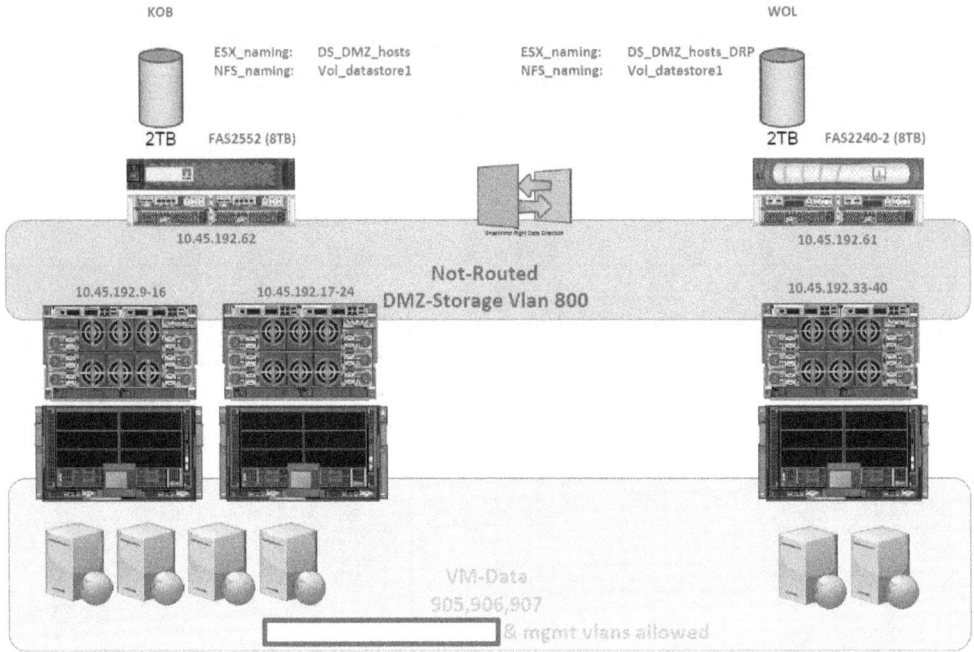

### 2.2.9 Monitoring

SolarWinds is used to monitor the availability and usage of the WAN, LAN and DMZ infrastructure.
SolarWinds Virtualization manager is used to monitor usage and availability of the DMZ virtual environment.

### 2.2.10 InFlight projects

Integration of Citrix NetScaler
Setup of a Mobile Device Management Tool – Citrix XenMobile

### 2.2.11 Information Library

| Context | File name | Content |
|---|---|---|
| **DC_NETWORK_Topology_Detail.vsd** | DC_NETWORK_Topology_Detail.vsd | View on DC connectivity + LAN |
| **IP VLAN Configuration** | IP-VLAN configuration.xlsx | VLAN Information |
| **CUS LAN** | CUS LAN Core 5.1.vsd | View on DC Lan Setup |
| **DMZ Application list** | DMZ Server overview.xlsx | Overview of DMZ servers |
| **Internet Gateway** | Internet gateway layout.jpg | Design of Internet Gateway |

## 2.3 LOT 2 – SAP & STORAGE

### 2.3.1.1 Functionality

SAP has been chosen as strategic application platform for the Customer Group. The following processes are covered by SAP applications:

Order to Cash
Procure to Pay
Finance and Controlling
Capex Follow-up
Production Planning
Plant Maintenance
CRM (Customer Relationship Management + Mobile)
Business Intelligence (reporting)
SuPM : Sustainability Performance Mgt

The SAP applications are business critical and need to be available 24x7.

### 2.3.2 SAP Systems

The SAP systems run on 2 different platforms: iSeries and AIX.
The ERP, CRM and BW system run on the iSeries platform. As the BW database has been migrated to HANA, only the BW application server runs on the AS400 platform. The iSeries infrastructure is setup in high availability mode over the 2 datacenters using MIMIX for asynchronous software replication and the hardware in the Datacenter in Colocation is also used to run the development and test systems.

The SAP SCM (Supply Chain Management), XI (Interfacing), GRC (NotaFiscal for Brazil) run on AIX and uses NetApp as storage. The replication to the Datacenter in Colocation is done using Snapshot and mirroring features from NetApp. The hardware in Colocation is also used to run the development and test systems.

For Master Data Management an IBM solution has been installed on the AIX environment. The replication to the Datacenter in Colocation is done using Snapshot and mirroring features from NetApp. The hardware in Colocation is also used to run the development and test systems.

We are in process of setting up a Business Objects environment on Windows which will run on a VMware platform.

### 2.3.2.1 System copy

Every year we do a system copy from the production systems to the quality systems. For the ERP system, we have recently installed a new concept to reduce the amount of data (currently this is set to 13 months) which is copied to the quality system.

Every week, we do a system copy from the production ERP system to a simulation system. Also here, DSM is used to limit the data to 6 months. This system is used to test transports as well as to simulate certain issues from the production system.

### 2.3.2.2 Transports

**iSeries**

Transport directory is owned by BEMANI11 partition and is linked to by other systems using QFileSvr.400. For example in BEMANI01 (development) directory /QFileSvr.400/BEMANI11 will provide the link to the owner of transport directory. SAP accesses the OS-files by means of the ADM user. Each SAP system has its own ADM-user. For example SD1 has user SD1ADM which must exist also on BEMANI11 with the same password.

All our SAP-systems are 3-tier systems consisting of Development, Quality and Production. Transports can only be generated in the Development system. Once these are released they are placed on the import queue of the Quality environment. An import-all job runs here with a 15 minute interval. Once the import in Quality is done the transport is placed on the production queue where it will remain until the import is requested by the functional responsible. The process of requesting the import into Production is handled in the Eusap Call-database, where the transport number is placed on a list which is picked up in the evening by people from service desk, who put them in a job to be imported next morning at 05:30 hrs CET.

**pSeries (AIX)**

On AIX, the transport directory is owned by BEMANP02 and is linked by the other systems. Transport flow is similar as for iSeries.

**Upgrades**

Usually we schedule an upgrade every 2 years. Upgrades are in close collaboration with the SAP functional teams. For the larger systems, a sandbox upgrade is performed first and an emergency system is created to still allow changes while the SAP landscape (Development-Quality-Production) is being upgraded. Upgrade of production systems is always executed during weekends.

### 2.3.2.3 Number of users

We have over 5000 named users in our SAP landscape.
The number of concurrent users in our main system (ERP) is shown in the graph below as an example:

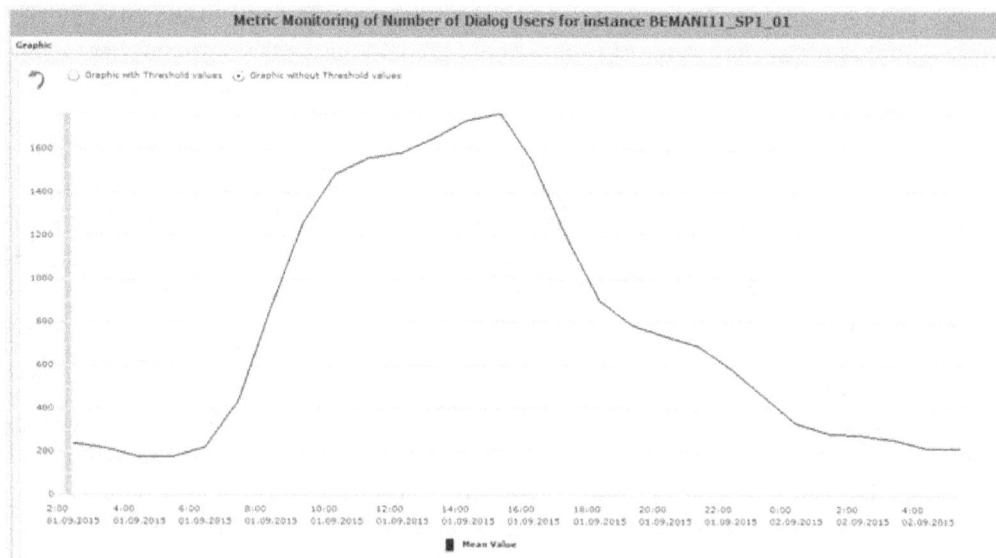

### 2.3.3 SAP Storage

#### 2.3.3.1 iSeries

**CUS - Datacenter**

The partitions on the P770 have dedicated 8 Gbps FC HBA's directly connected to an IBM DS8870 storage system, with 240 300GB 15K SAS disks.

This is the overview of the LUN's per system:

| System | #HBA's | LUN size | #LUN's |
| --- | --- | --- | --- |
| BEMANI11 | 12 | 65,7 GiB | 202 |
| BEMANI12 | 4 | 131,4 GiB | 91 |
| BEMANI13 | 4 | 65,7 GiB | 64 |
| BEMANI14 | 2 | 131,4 GiB | 40 |

For each LUN there are 2 paths. The IBM DS8000 Storage Manager is used to manage the storage system. This is a web based client.

**Colocation - Datacenter**

The partitions on the P770 have dedicated 4/8 Gbps FC HBA's connected to 2 8Gbps SAN switches connecting to an IBM DS8700 storage system, with 224 300GB 15K SAS disks and 32 450 GB 15K disks.

This is the overview of the LUN's per system:

| System | #HBA's | LUN size | #LUN's | #paths/lun |
|--------|--------|----------|--------|------------|
| BEMANI11HA | 8 | 65,7 GiB | 171 | 2 |
| BEMANI12HA | 4 | 65,7 GiB | 154 | 2 |
| BEMANI13HA | 4 | 65,7 GiB | 63 | 2 |
| BEMANI01 | 4 | 65,7 GiB | 71 | 2 |
| BEMANI02 | 12 | 131,4+65,7 GiB | 188 | 2 |
| BEMANI03 | 1 | 131,4 GiB | 36 | 1 |
| BEMANI04 | 1 | 131,4 GiB | 8 | 1 |

Each of the paths goes to a SAN switch, which than goes to the storage system.

The IBM System Storage DS8000 Storage Manager is used to manage the storage system. This is a web based client.

**Backup**

On the iSeries, BRMS is used to backup the systems to IBM 3592-E07 tape drives. These are FC attached drives. We have 4 of them in a 3584-022 library in Colocation and 1 standalone in CUS share by the production systems. We use 10 tapes in CUS and 115 in Colocation. All of the systems can access all the drives, but will only use 1 at the time.

The SAP database is backed up every day.

Once a week, a system backup is taken (savsys), except for the production systems (+HA) which are backed up after PTF installation. Once a month this backup becomes the monthly backup.

Weekly tapes expire after 31 days, daily tapes after 7 days and monthly tapes after 300 days.

### 2.3.3.2 SAP HANA

HANA database is synchronized to Colocation via the built-in HANA synchronization. Backups are taken via HANA Studio onto the central NetApp storage.

### 2.3.3.3 AIX Storage

The AIX systems use the central NetApp as storage system.

The Power/AIX infrastructure at is distributed over two datacenters, one in Location 1 and one in Colocation. Each datacenter contains a set of blades that are configured in such a way that

they provide a pool of completely virtualized computing resources on which the different workloads can be deployed as virtual machines (LPARs). These LPARs can be moved from one blade to another without service disruption using Live Partition Mobility.
The central point of control is the HMC of Colocation: a management console used to define, start/stop and move the LPARs.

This architecture offers two levels op protection
- When one blade would fail, or needs to be stopped for maintenance, the workload can be moved to another blade.
- If a large component on a site fails or if a complete site fails, workloads from one datacenter can be started in the other datacenter.

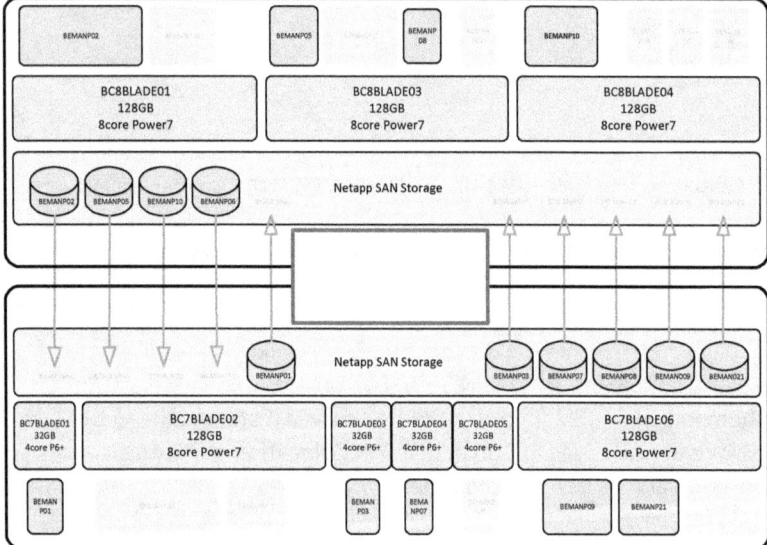

The central Storage system is a NetApp 8040 cluster with 2 nodes in CUS and another NetApp 8040 cluster with 2 nodes in Colocation. This infrastructure is the central storage which is used as NAS and as SAN. Each controller has 1 TB of cache.

In CUS one controller has 94 2TB 7K2 SATA disks attached and the other controller has 157 600GB 10K SAS disks. The controller with the SATA disks is used for CIFS and NFS access and the controller with the SAS disks is used for FC access. The NetApp volumes containing the CIFS/NFS data and the LUN's are snap mirrored to the controllers in Colocation.

In Colocation one controller has 24 2TB 7K2 SATA disks and 46 4TB 7K2 SATA disks attached and the other controller has 157 600GB 10K SAS disks. The controller with the SAS disks receives the

mirrors from the SAS-volumes and the controller with the SATA disks receives the mirrors from the SATA-volumes. Here is more capacity (on SATA) to store snapvaults.

For the backups of CIFS and FC, SnapCreator is used, which will put the application data in a consistent state and then take a NetApp snapshot, which is then mirrored to the mirror volume.

## CUS – Datacenter - AIX

This is the main datacenter housing Bladecenter BC8. The blades service the production workloads.

| Location 1 | | | | |
|---|---|---|---|---|
| **Blade** | **cores (#)** | **mem (GB)** | **Suggested workload** | |
| | | | **non-DRP** | **DRP** |
| BC8BLADE01 | 8 | 128 | BEMANP02 (PROD-APO) | BEMANP02 BEMANP09_drp |
| BC8BLADE03 | 8 | 128 | BEMANP05 BEMANP06 (PROD-XI, PROD-BO ) | BEMANP05 BEMANP06 BEMANP21_drp BEMANP01_drp |
| BC8BLADE04 | 8 | 128 | BEMANP10 (PROD-MDM) | BEMANP10, BEMANP01_drp BEMANP07_drp BEMANP08_drp |

Storage is provided by a NetApp storage server and data is mirrored asynchronously to the datacenter in Colocation.

## Colocation – Datacenter - AIX

This datacenter houses Bladecenter BC7. The blades service
- development and QA workloads
- production workloads during DRP

| Colocation | | | | |
|---|---|---|---|---|
| **Blade** | **cores (#)** | **mem (GB)** | **Suggested Workload** | |
| | | | **non-DRP** | **DRP** |
| BC7BLADE01 | 4 | 32 | BEMANP01 (DEV-APO/XI) | BEMANP01 |
| BC7BLADE02 | 8 | 128 | | BEMANP02_drp, BEMANP05_drp |
| BC7BLADE03 | 4 | 32 | BEMANP03 (QA-XI, QA-APO) | BEMANP03 |
| BC7BLADE04 | 4 | 32 | BEMANP07 (NIM, ILMT) | BEMANP07 |
| BC7BLADE05 | 4 | 32 | | BEMANP06_drp |
| BC7BLADE06 | 8 | 128 | BEMANP09 BEMANP21 (QA MDM) (DEV-MDM) | BEMANP09 BEMANP21 |

## Blade Configuration

Each blade contains a single VIO server. The VIO server offers virtual fiber channel and virtual network interfaces to the AIX LPARs.

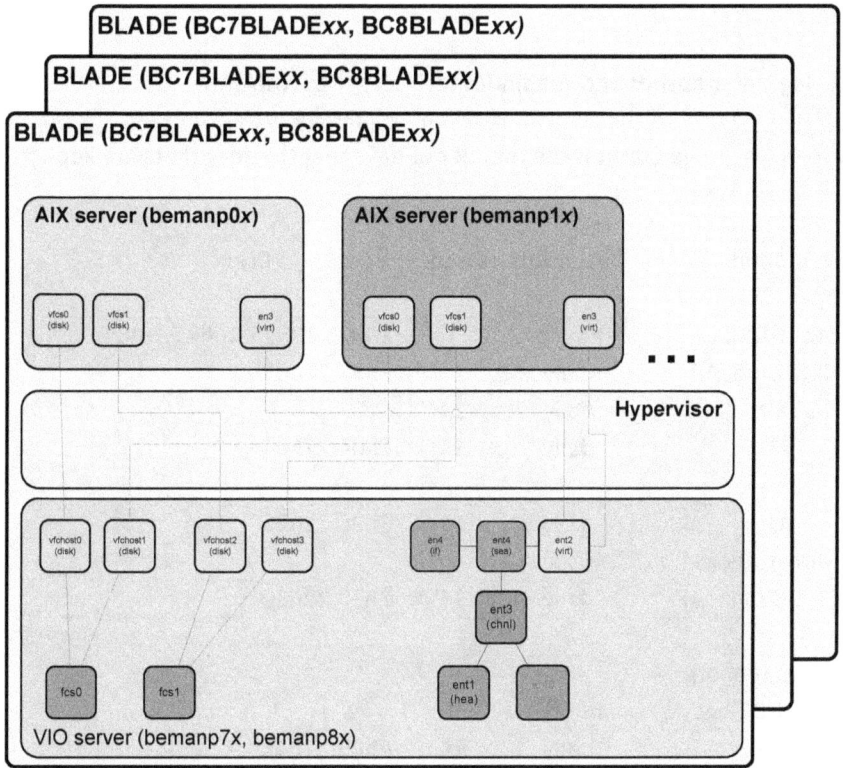

## Storage configuration

All LPARs, including the VIO server boot from the SAN. The two FC HBA's of the blades are assigned to the VIO server.

The other LPARs use two virtual fiber channel adapters that are mapped to the two HBA's in the VIO server (NPIV).

Zoning on the SAN switches is port based. LUNs are assigned to the corresponding WWPN using LUN masking on the NetApp storage server:

- Two physical WWPNs for the VIO servers
- Four virtual WWPNs for the LPARs.

**Backup**

**Backups are taken using the snapshot and snapvault technology of NetApp.**
The following table is an overview of the snapshots taken by SnapCreator. Cron gives the time of day the snapshots are taken, while Count is the number of days that the snapshots are kept. Snapshots older than Count are deleted.

| Profile | Config | Retention | Count | Cron |
|---|---|---|---|---|
| **bemanp01** | | | | |
| | MirrorDBlogs | hourly | 1 | 10h25,13h25,16h25,19h25 |
| | AllDB2andSAP | daily | 14 | 7h05,22h05 |
| | XD1 | daily | 14 | |
| | LD1 | daily | 14 | 7h04,22h04 |
| **bemanp02** | | | | |
| | LP1 | daily | 14 | 7h25,22h25 |
| | MirrorDBlogs | | | |
| | AP1_DB2andSAP | daily | 14 | 7h30,22h30 |
| **bemanp03** | | | | |
| | db2AQ1db_org | - | - | |
| | MirrorDBlogs | hourly | 1 | |
| | LQ1 | daily | 14 | 6h49,21h49 |
| | AllDB2andSAP | daily | 14 | 6h50, 21h50 |
| | db2AQ1dbeverything | hourly | 72 | |
| **bemanp05** | | | | |
| | XP1_DB2andSAP | daily | 14 | 6h50,21h50 |
| **bemanp08** | | | | |
| | GP1_DB2andSAP | daily | 14 | 6h50,21h50 |
| **bemanp09N** | | | | |
| | FullMQ1 | daily | 14 | 07h15,20h15 |
| **bemanp10** | | | | |
| | FullMP1 | daily | 14 | 07h25,20h25 |
| **bemanp18** | | | | |
| | LDS | daily | 14 | 4h59,17h59 |
| | ADS_DB2andSAP | daily | 14 | 5h00,18h00 |
| **bemanp21** | | | | |
| | FullMD1 | daily | 14 | 20h12,7h12 |

For all the AIX systems, also a snapvault backup copy is created at Colocation, once a day. The retention is 56 days.

### 2.3.4 Monitoring

SAP solution manager is installed to monitor availability and usage of SAP production systems.

### 2.3.5  Support Agreements

| Items | Provider | Hours |
|---|---|---|
| **iSeries (Power, DS systems)** | .../IBM | 24x7 |
| **NetApp** | ... | 24x7 |
| **AIX systems (PowerBlades and Chassis via IBM)** | IBM/... | 24x7 |

### 2.3.6  Performance

1. SAP ERP performance via collection of ST03/STAD statistics on the ERP and CRM production system

2. System i Navigator for iSeries systems

### 2.3.7  Information Library

| Context | File name | Content |
|---|---|---|
| **Hardware inventory** | SAP Inventory.xlsx | Details on Hardware |
| **Software inventory** | SAP Inventory.xlsx | Details on SAP applications |
| **Early Watch Reports** | Different EWA per system | |
| **SAP Landscape** | System Landscape 770 (p7) + Blades.pdf | |
| **Performance Stats** | | |
| **AIX Partitions** | Aix partitions.xlsx | Technical info on AIX systems |
| **SAP Monitoring** | SAP Monitoring.xlsx | Description of monitoring parameters |

## 2.4 LOT 3 – VMware – Windows environment

### 2.4.1 Platform descriptions: Virtual Environment

We have several VMware cluster environments active in the datacenters:

Cluster CUS – Used for most applications
Cluster DMZ – Used for all applications in the DMZ (see chapter Datacenter Connectivity DMZ)
Cluster Email – Used for the large Email servers

### 2.4.2 Cluster CUS

The VMware cluster CUS consists of 7 hosts in the datacenter CUS and uses storage from NetApp. The Virtual Machines are mirrored to the datacenter in Colocation using NetApp snapshot and mirroring technology.

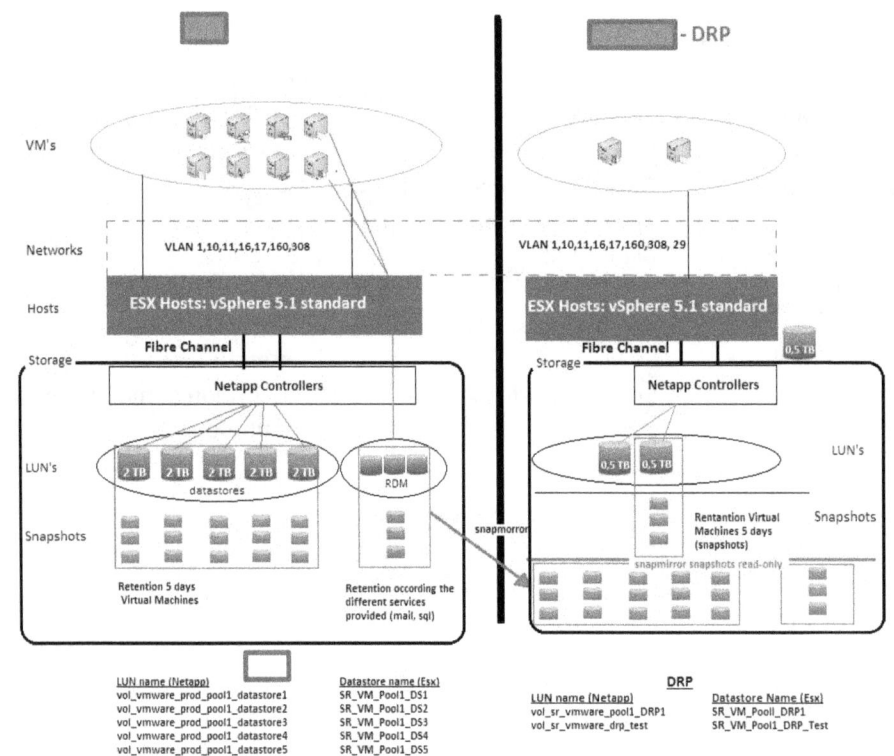

Due to the diversity of applications that is hosted on this VM Cluster, the criticality varies from business critical to less critical. Therefore the availability of the VM platform has to be 24/7.

### 2.4.3    Cluster Email

A VMware cluster is being built to host the larger Email servers. The VMware cluster Email consists of 3 hosts in the datacenter CUS and uses storage from NetApp. The Virtual Machines are mirrored to the datacenter in Colocation using NetApp snapshot and mirroring technology.

| General | |
| --- | --- |
| vSphere DRS: | Off |
| vSphere HA: | On |
| VMware EVC Mode: | Disabled |
| Total CPU Resources: | 112 GHz |
| Total Memory: | 239,72 GB |
| Total Storage: | 500,00 GB |
| Number of Hosts: | 3 |
| Total Processors: | 36 |
| Number of Datastore Clusters: | 0 |
| Total Datastores: | 1 |
| Virtual Machines and Templates: | 5 |
| Total Migrations using vMotion: | 0 |

| General | |
| --- | --- |
| Manufacturer: | HP |
| Model: | ProLiant BL460c Gen9 |
| CPU Cores: | 12 CPUs x 3,396 GHz |
| Processor Type: | Intel(R) Xeon(R) CPU E5-2643 v3 @ 3.40GHz |
| License: | VMware vSphere 5 Standard - Licensed for 2 physical CPU... |
| Processor Sockets: | 2 |
| Cores per Socket: | 6 |
| Logical Processors: | 24 |
| Hyperthreading: | Active |
| Number of NICs: | 2 |
| State: | Connected |
| Virtual Machines and Templates: | 2 |
| vMotion Enabled: | Yes |
| VMware EVC Mode: | Disabled |
| vSphere HA State | Running (Master) |
| Host Configured for FT: | No |
| Active Tasks: | |

### 2.4.4    Monitoring

SolarWinds Virtualization Manager is used to monitor usage, performance and health of the VMware environment.

### 2.4.5    Support Agreements

| Items | Provider | Hours |
| --- | --- | --- |
| **VMware** | ProAct | 24x7 |
| **Storage NetApp** | ProAct | 24x7 |
| **Hosts** | HP | 24x7 |

### 2.4.6    InFlight projects

Upgrade hosts to VMware vSphere standard 5.5

### 2.4.7 Platform Description File Server

Centrally we provide a file server, which is used by all Citrix users to store data as well as the Citrix profiles, mail files, etc…
The performance of the file server is heavily linked to the user experience for the Citrix users.

**Retention**
The file server is hosted on the central NetApp in the datacenter in CUS and mirrored to the datacenter in Colocation using snapshot technology.
End users are able to restore files using the previous versions functionality up to 8 weeks ago.

Using Snapvault technology, backups are taken up to 12 weeks but a restore requires the invention of the central IT team.

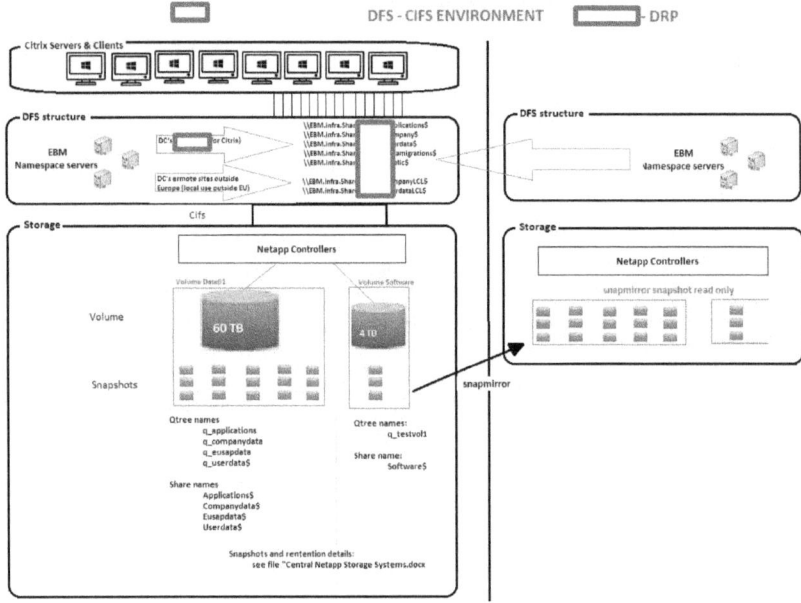

**CIFS NetApp volumes in CUS**

| Name | Aggregate | Status | Thin Provisioned | % Used | Available Space | Total Space | Storage Efficie... |
|------|-----------|--------|------------------|--------|-----------------|-------------|--------------------|
| BackupSQL01 | aggr0_spso008_sata | online | Yes | 18 | 4.91 TB | 6 TB | Disabled |
| DocRoom01 | aggr0_spso008_sata | online | Yes | 39 | 622.37 GB | 1 TB | Enabled |
| LotusNotesDAOS02 | aggr0_spso008_sata | online | Yes | 48 | 5.17 TB | 10 TB | Enabled |
| SPFW0110_root | aggr0_spso008_sata | online | No | 5 | 972.44 MB | 1 GB | Disabled |
| audit | aggr0_spso008_sata | online | Yes | 5 | 94.47 GB | 100 GB | Disabled |
| backup_saphana_dev_mirror | aggr0_spso008_sata | online | Yes | 9 | 4.5 TB | 5 TB | Enabled |
| backup_saphana_prd | aggr0_spso008_sata | online | Yes | 5 | 9.5 TB | 10 TB | Enabled |
| backup_saphana_qa_mirror | aggr0_spso008_sata | online | Yes | 13 | 4.33 TB | 5 TB | Enabled |
| backups01 | aggr0_spso008_sata | online | Yes | 9 | 2.71 TB | 3 TB | Enabled |
| data01 | aggr0_spso008_sata | online | Yes | 73 | 15.98 TB | 60 TB | Enabled |
| sharefile01 | aggr0_spso008_sata | online | Yes | 60 | 402.17 GB | 1 TB | Enabled |
| software | aggr0_spso008_sata | online | Yes | 36 | 2.54 TB | 4 TB | Enabled |
| vol_as400_test | aggr0_spso008_sata | online | Yes | 5 | 475 GB | 500 GB | Disabled |
| vol_av_test | aggr0_spso008_sata | online | Yes | 16 | 857.36 GB | 1 TB | Disabled |
| vol_test | aggr0_spso008_sata | online | Yes | 12 | 67.66 GB | 100 GB | Enabled |

## CIFS NetApp volumes in Colocation, including snapvault volumes

| Name | Aggregate | Status | Thin Provisioned | % Us... | Available Space | Total Sp... | Storage Efficiency |
|------|-----------|--------|------------------|---------|-----------------|-------------|--------------------|
| BackupSQL01_mirror | aggr1_spso012_sata | online | Yes | 24 | 4.54 TB | 6 TB | Disabled |
| LotusNotesDAOS02_mirror | aggr1_spso012_sata | online | Yes | 48 | 5.17 TB | 10 TB | Enabled |
| SRFW0110_root | aggr0_spso012_sata | online | No | 5 | 972.02 MB | 1 GB | Disabled |
| SRFW0110_root_mirror | aggr0_spso013_sas | online | No | 5 | 972.02 MB | 1 GB | Disabled |
| backup_saphana_dev | aggr1_spso012_sata | online | Yes | 10 | 4.49 TB | 5 TB | Enabled |
| backup_saphana_prd_mirror | aggr0_spso012_sata | online | Yes | 5 | 9.5 TB | 10 TB | Enabled |
| backup_saphana_qa | aggr0_spso012_sata | online | Yes | 14 | 4.28 TB | 5 TB | Enabled |
| backups01_mirror | aggr0_spso012_sata | online | Yes | 9 | 2.71 TB | 3 TB | Enabled |
| data01_mirror | aggr0_spso012_sata | online | Yes | 73 | 16.03 TB | 60 TB | Enabled |
| sharefile01_mirror | aggr1_spso012_sata | online | Yes | 60 | 402.05 GB | 1 TB | Enabled |
| vol_as400_test | aggr0_spso012_sata | online | Yes | 5 | 95 GB | 100 GB | Disabled |
| vol_bemanp10_rootvg_backup | aggr0_spso012_sata | online | Yes | 2 | 292.96 GB | 300 GB | Enabled |
| vol_bemanp10_vgmp1dat01_backup | aggr0_spso012_sata | online | Yes | 2 | 292.75 GB | 300 GB | Enabled |
| vol_bemanp10_vgmp1dat02_backup | aggr0_spso012_sata | online | Yes | 2 | 292.75 GB | 300 GB | Enabled |
| vol_bemanp10_vgmp1dat03_backup | aggr0_spso012_sata | online | Yes | 2 | 292.78 GB | 300 GB | Enabled |
| vol_bemanp10_vgmp1dat04_backup | aggr0_spso012_sata | online | Yes | 2 | 292.78 GB | 300 GB | Enabled |
| vol_bemanp10_vgmp1dat1_backup | aggr0_spso012_sata | online | Yes | 3 | 288.55 GB | 300 GB | Enabled |
| vol_bemanp10_vgmp1log_backup | aggr0_spso012_sata | online | Yes | 10 | 896.79 GB | 1000 GB | Enabled |
| vol_bemanp10_vgmp1suf1_backup | aggr0_spso012_sata | online | Yes | 9 | 271.17 GB | 300 GB | Enabled |
| vol_bemanp10_vgmp1wmb1_backup | aggr0_spso012_sata | online | Yes | 9 | 45.41 GB | 50 GB | Enabled |
| vol_spmw0001_g_backup | aggr0_spso012_sata | online | Yes | 96 | 83.38 GB | 2.1 TB | Enabled |
| vol_spmw0001_i_backup | aggr1_spso012_sata | online | Yes | 34 | 1.32 TB | 2 TB | Enabled |
| vol_spmw0003_g_backup | aggr1_spso012_sata | online | Yes | 36 | 1.9 TB | 3 TB | Enabled |
| vol_spmw0008_E_backup | aggr0_spso012_sata | online | Yes | 75 | 3.47 TB | 14 TB | Enabled |
| vol_spmw0009_E_backup | aggr1_spso012_sata | online | Yes | 61 | 5.42 TB | 14 TB | Enabled |
| vol_spmw0010_E_backup | aggr0_spso012_sata | online | Yes | 0 | 6 TB | 6 TB | Enabled |
| vol_spmw0011_E_backup | aggr0_spso012_sata | online | Yes | 0 | 6 TB | 6 TB | Enabled |

**Backup retentions**

The following table is an overview of the snapshots taken by SnapCreator. Cron gives the time of day the snapshots are taken, while Count is the number of days that the snapshots are kept. Snapshots older than Count are deleted.

| Profile | Config | Retention | Count | Cron |
|---|---|---|---|---|
| smb_shares | | | | |
| | ShareFile01 | hourly | 36 | 5h00, 7h00,9h00,11h00,13h00,15h00, 17h00,19h00,21h00 |
| | software | daily | 6 | 20h00 |
| | docroom01 | daily | 2 | 4h00 |
| | data01 | hourly | 36 | 5h00, 7h00,9h00,11h00,13h00,15h00, 17h00,19h00,21h00 |
| | backupsql01 | daily | 4 | 6h15 |
| | data01_daily | daily | 28 | 23h00 |
| | ShareFile01_daily | daily | 28 | 23h00 |
| | backups01 | daily | 7 | 1h30 |
| | | weekly | 4 | 1h30 |
| | data01_weekly | weekly | 10 | 23h00 |

For all the CIFS (Data01 and ShareFile) also a snapvault backup copy is created at Colocation, once a day. The retention is 84 (12 weeks).

**CIFS Antivirus**

The file server is protected with an Anti-virus solution from McAfee. 4 **physical servers** are connected to the NetApp in the datacenter of CUS to scan files before they are written on the file server.

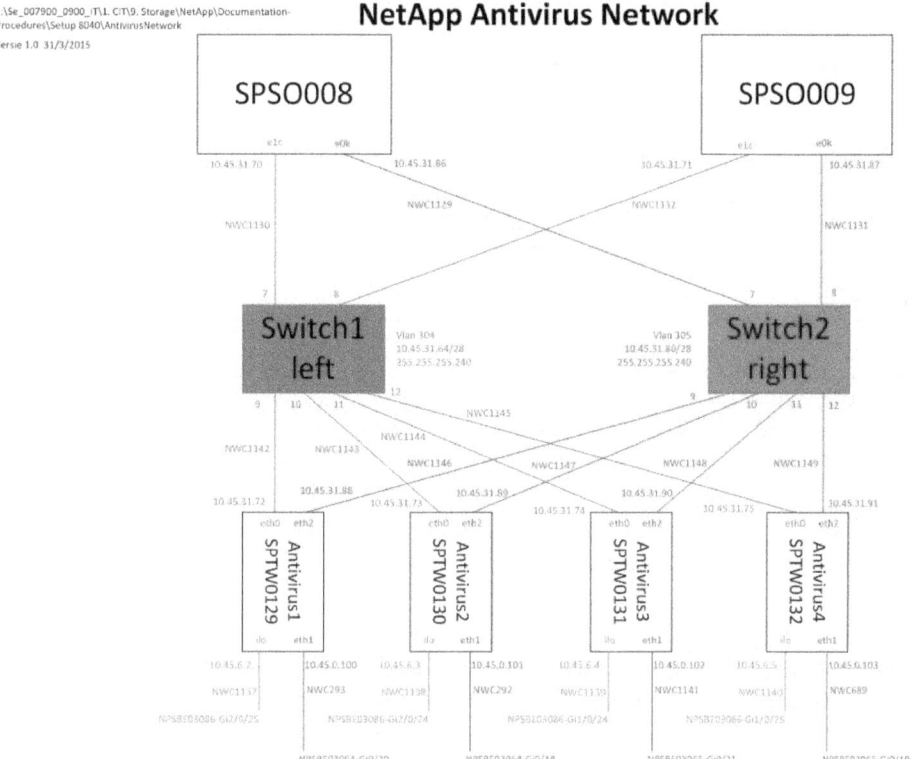

### 2.4.8 Data structure within CUSTOMER

Due to added interactions with sister companies and the need to share data between them (logistics, marketing shared service centers in accounting, ...), it's necessary to align the data structure and create a system that makes the sharing of information within Citrix possible. When you open a Windows Explorer in Citrix you will see several network drives.

| Drive | Usage of Drive |
|-------|----------------|
| E: | Customer Companies |
| F: | Company Data |
| G: | Cross Company Data |
| H: | Home Drive |
| O: | Application Data |
| P: | Project Folder |
| T: | Local Company Data |
| V: | Local C-drive (optional) |
| X: | Link to old shared folders |

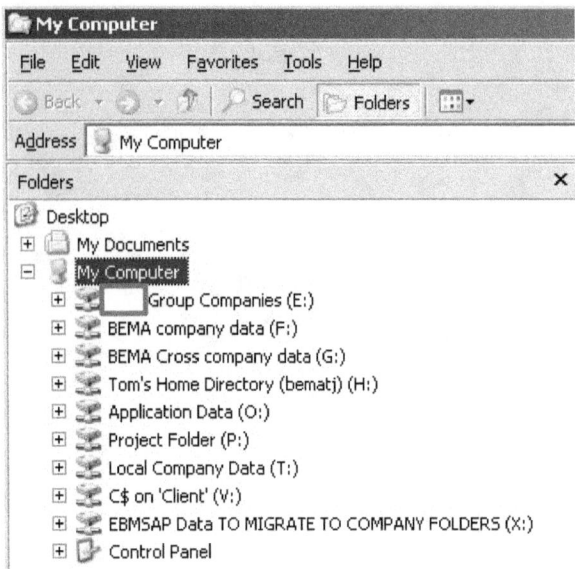

The new folder structure allows mappings that are the same for every user; this means that everybody have the same network drives wherever they logon.
All these drives are automatically saved with a backup system.

### The Company Drive F: Layout
The F: drive is our company data folder. Each company will have the same data structure.
There are always 13 folders inside related to the most commons departments.
Here you will find company data, only accessible for the employees from that company.

### The Company Drive F: Security
2 Layers of security are applied in the folder structure, all files & folders below will inherit the security from their parent.
When needed, you can put folders between the 2 security layers for customization.
Each security layer will have 3 security groups (Read Only, Read/Write and Admin).
Each folder with separate security rights will have a sequential number which will reflect accordingly in the naming convention of the security groups.

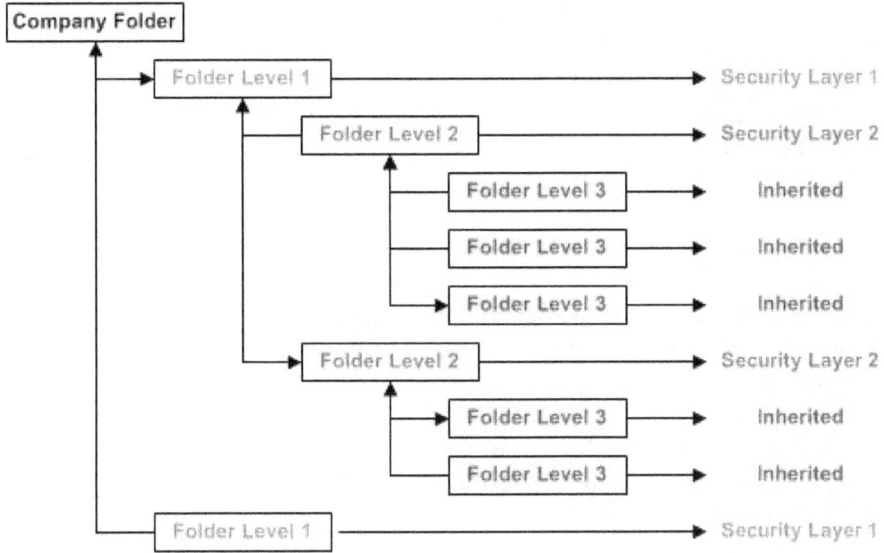

In picture below Folder Level 2 could be different business units within a country/company.

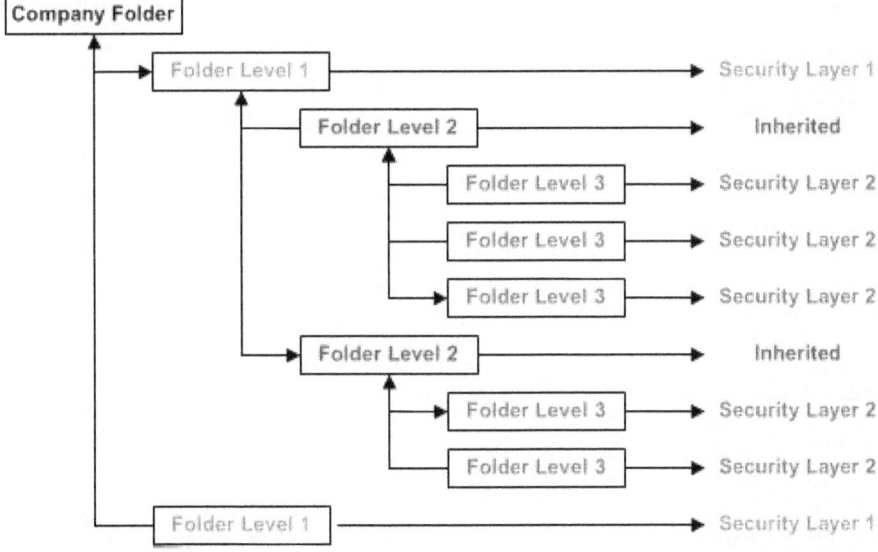

### The Cross-Company Drive G:

Whenever there is a need to share data across different companies this data can be added on the G-Drive. The G-drive allows the users to save data that can be accessed by a colleague from another company.

The root structure of the G-drive cannot be changed and is an exact copy of the F-drive. The folder structure below is a decision to be made by the company. An idea is to keep those folders exactly the same as on the F-Drive.

When a file or files are added to the G-drive any other person of any other company can access this data through the E: Drive. The E-Drive is sorted on the company code.

The requirements to share data with other companies are:

Both companies need to be on the Shared Customer Domain
The users of both companies need access to Citrix
Both users need access rights to that location and data

Access rights are given by the Local IT of that specific company.

**Example:**
User A from company AA with company code 8888 adds a file in the root folder
G:\Se_0088888_900_IT for user B of company BB with company code 6666.
User A gives read access rights to user B on folder G:\Se_0088888_900_IT.
Now user B can access the data only through the E: Drive because user B its G:-drive is
G:\Se_006666_900_IT.

The G: drive is the cross company data folder of your company.
The G: drive can host data for other users within the CUSTOMER Group.
You will see the same data structure like the F: drive.

**The Companies Drive E:**
This drive gives you access to all the cross company drives.

- Etex Group Companies (E:)
  - Se_000000_EUBM
  - Se_000010_BEEG
  - Se_000030_BEEL
  - Se_000040_BEWE
  - Se_000050_BEES
  - Se_001200_DEET
  - Se_001220_DEEH
  - Se_001300_DEEF
  - Se_001500_DEWF
  - Se_001600_DEPF
  - Se_001700_DECR
  - Se_001840_AUCR
  - Se_001880_PLCR
  - Se_002100_PLEU
  - Se_002200_LTET

**The Home Drive H:**
The H: drive is a personal "home" directory. Only the user can read & write on this location and
only the user has access to the home drive.
The home drive is only for personal work-related data. Policy: The user must not use the home
drive for personal data like family photos, music, video's,...
All work related data that isn't confidential or confidential within the department, should be put
in the F: drive instead of in the H: drive.

**The Application Drive O:**
This location hosts data used by applications which needs a centralized location for multiuser
support.

**The Project Drive P:**
Is also called the EUSAP drive. Used by SAP application teams for project related data.

**Migration Drive T: (Temporary)**
This drive is used for the migration in Germany and will eventually disappear.

**Migration Drive X: (Temporary)**
Drive which holds data from the previous environment (EBMSAP). Also the X-drive contains folder access to fileserver of companies that are in a migration process.

**"Local" Drive V:**
Drive that maps your local C: drive to your Citrix session.
To remove the local drive, add the user to security group **EBM\GG00CITRIX-NOLOCDRIVESxP**

**"Local" Drive USB Device**
Drive that maps your local USB disk to you Citrix session. The letter used to mount the drive depends on the letter used on the local computer.
To add the local USB device, add the user to security group **EBM\GG00CITRIX-USBACCESSxP**

**"Local" CD/DVD-ROM Drive**
Drive that maps your local CD/DVD drive to you Citrix session. The letter used to mount the drive depends on the letter used on the local computer.
To add the local CD/DVD drive, add the user to security group **EBM\GG00CITRIX-CDROMACCESSxP**

In general the use of local drives within Citrix is discouraged as it has an impact on performance and backups are not taken by central IT of local drives.
Citrix ShareFile is also available to synchronize data between laptops, mobile devices and the central file server.

### 2.4.9    Platform Description ShareFile

ShareFile is a tool that enables users to store and sync files on any device at any place. This enables users to use their Office files while offline and because these files are automatically synced to the shared datacenter in CUS. These files are also available within the XenApp Environment. Because the files are stored in the Central datacenter these file are also backed-up. These are a lot of additional features that are made available with this tool.

For the setup at Customer we only make use of on premise storage zones. This gives users the complete control over our own data (backup/recovery/management) and makes it possible to show the data in the user's XenApp session. Meaning that all data that is synced through ShareFile from the local PC is also available in the users XenApp session. Changes made to these files are also synced back to the local client.

## Citrix

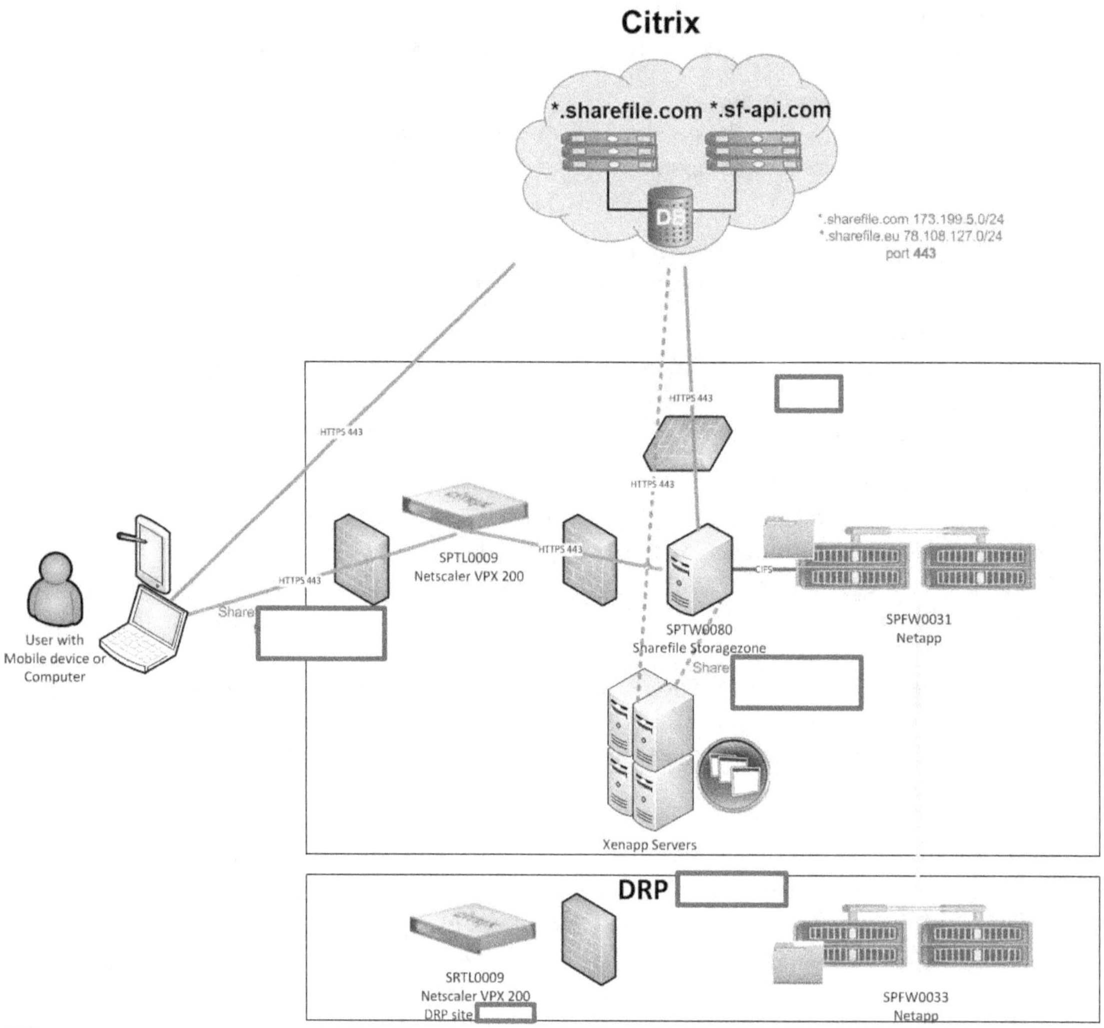

**\*.sharefile.com \*.sf-api.com**

*.sharefile.com 173.199.5.0/24
*.sharefile.eu 78.108.127.0/24
port **443**

HTTPS 443

HTTPS 443

HTTPS 443

HTTPS 443

HTTPS 443

SPTL0009
Netscaler VPX 200

SPTW0080
Sharefile Storagezone

SPFW0031
Netapp

User with
Mobile device or
Computer

Xenapp Servers

## DRP

SRTL0009
Netscaler VPX 200
DRP site

SPFW0033
Netapp

CUS setup

As mentioned before the Citrix Control Plane is hosted in a Citrix datacenter. Therefore a lot of the used components need a direct connection to the control plane on the internet. These links are shown as a blue line. In order to use an on premise storage zone at least 3 components are required.

### NetScaler

The NetScaler acts as a reversed proxy to secure the traffic coming from the internet. This device handles all internet traffic for the storage zone SharefSZ01.Customer.be and acts as a buffer to the Storage zone Controller securing this component from unwanted access from the internet. There are 2 NetScalers setup working in HA mode. This means that one NetScaler SPTL0009 is an active node and one SRTL0009 is the passive node. If for some reason the active node fails the passive mode will take over without any loss in traffic.
These NetScalers are not only used for ShareFile.

### ShareFile StorageZone Controller

The storage zone controller is responsible for handling the file request for a storage zone, SharefSZ01 for CUS. The Storage zone controller makes use from IIS and needs a Cifs share to store all file data. If a file request is issued by the client, request is passed through to the Storage Zone Controler by the NetScaler. The controller will get the correct file from the Cifs share and pass it to the client via the NetScaler.

The Storage zone controller used in CUS is named SPTW0080, this server is hosted on the VMware environment. In a later phase a second Storage zone controller will be added to the storage zone. At the moment the high availability of this server is guaranteed sufficient by the VMware environment.

**Cifs Share NetApp**

As mentioned in the controller part, the actual data is stored in a Cifs share. This cifs share can be offered by any file server. In CUS this share is hosted on the NetApp server. This way the ShareFile data can take advantage of NetApp features like snapshotting and more importantly D-Dub. Because the data on the NetApp is automatically replicated to the NetApp appliance in Colocation this data is save from a NetApp crash and will be available during a disaster recovery.

**Flow of the data**

In order to give an idea how ShareFile functions, see the flowchart below that describes the process that is followed when downloading a file from ShareFile.

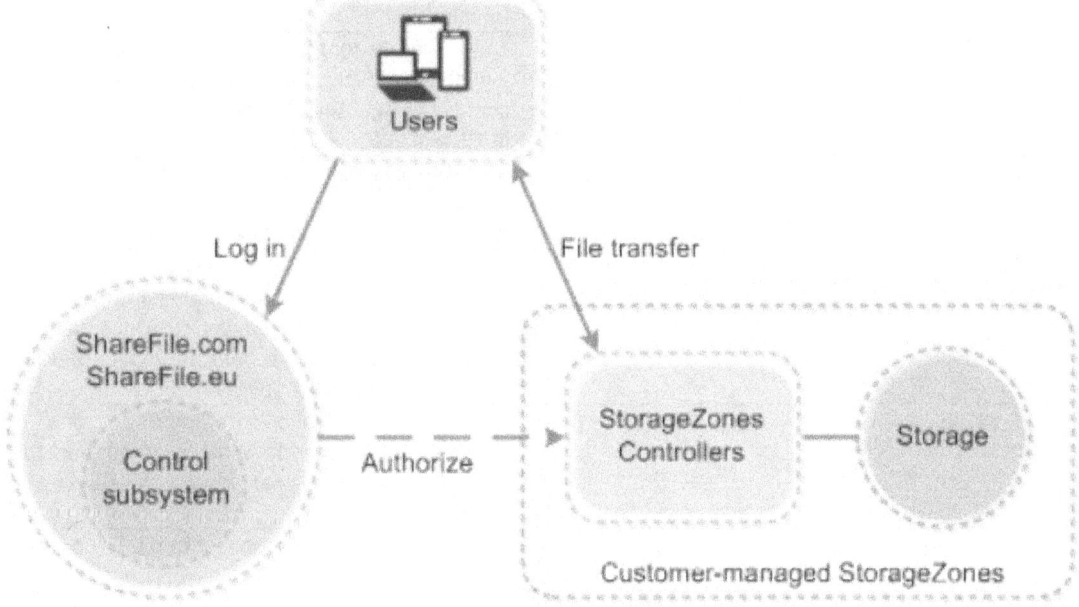

Flow

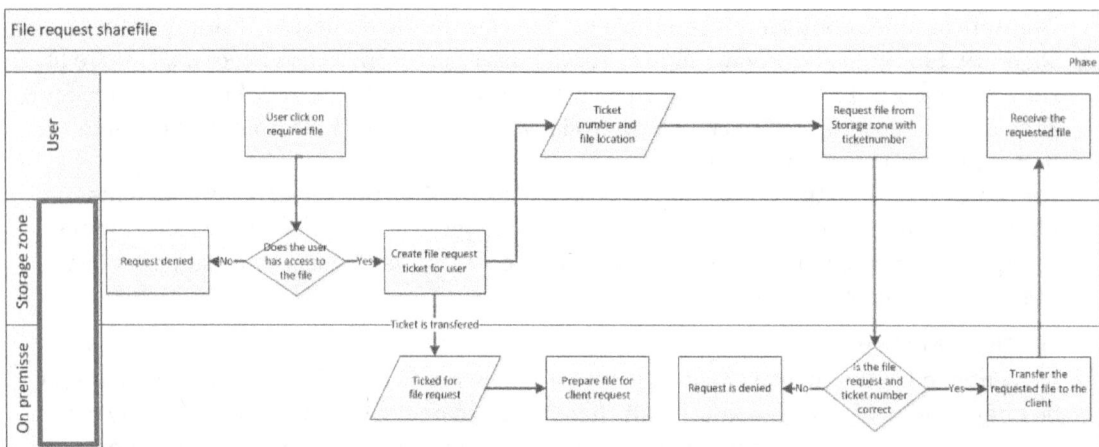

Flow file request

## Technical components

This chapter describes all components used in the ShareFile setup:

| Customer.ShareFile.eu or Customer.ShareFile.eu | This is the url to login to the Control plane. All clients need access to this url or ShareFile will not function. |
|---|---|
| SharefSZ01.Customer.be | This is the URL for the CUS storage zone. There are 2 IP defined for this URL depending where you connect you get the external IP xxx or the internal IP 10.45.0.140 |
| SPTL0009 | Main NetScaler VPX 200 responsible for handling the SharefSZ01.Customer.be traffic coming from the internet. |
| SRTL0009 | Backup NetScaler VPX 200 this NetScaler is setup as the backup NetScaler. If SPTL0009 fails this system will take over without a down time. Once the DMZ in the DRP site is ready this virtual appliance will move to the DRP DMZ environment. |
| SPTW0080 | This is the storage zone controller for the zone SharefSZ01.Customer.be. When users connect to this URL from the internal network this server will respond immediately. |
| SPFW0031 | This is the alias used to connect to the Cifs of the NetApp in CUS |
| SPFW0033 | This is the alias used to connect to the Cifs of the NetApp in DRP Colocation |

### 2.4.10 Platform description Microsoft SQL

Centrally we have installed a Microsoft SQL environment that hosts SQL Databases for several central applications.
The SQL databases are running on different servers in the datacenter CUS and use the central

NetApp storage. Transactional Log shipping is used to synchronize the databases with the datacenter in Colocation.

**Backup plan**
All non-system databases are being backed up in the following way

- Transaction log (every 2h) for the databases working in FULL recovery mode
- Incremental backup taken every evening (from 12:00AM until 3:00AM) (except on Friday)
- Full backup taken every Friday (from 12:00AM until 3:00AM)

**Databases synchronization between datacenters (CUS and Colocation)**
All non-system protected databases hosted on a SQL2008R2 server located in CUS have their transaction logs shipped to their secondary database server every 60min.

All non-system protected databases hosted on a SQL2012SP2 server located in CUS are mirrored to their secondary database server (synchronous mirroring)

### 2.4.11 Platform description Active Directory

Customer Wintel infrastructure is based on a Windows 2008 R2 forest with 3 domains and is created in 2007. The idea was to create a domain structure that is easy to maintain, easy to share data across the different Customer companies, easy to set different securities and easy to consolidate new or other companies of the group. To be able to do this we needed to standardize the different parts of the Wintel infrastructure.
A forest was created because back then there was a need that each division (EBM, IBM and OpCo) would have their own domain. Meanwhile this is no longer required because the structure of the group changed. 1 domain should be sufficient to support all the Customer companies and business units. No plans are foreseen to consolidate into 1 domain as it would take too much effort to move from 3 domains into 1 domain.
The domain has several domain controllers in the datacenter in CUS and Colocation as well as in the different Customer Sites.

| Location | Number of Domain controllers |
|---|---|
| CUS (BE01-CUS) | 6 Physical + 1 virtual |
| Colocation (BE100-DRPsite) | 2 Physical + 1 virtual |
| Azure | 1 virtual (for testing) |
| Customer Sites | 57 (37 Physical and 20 virtual) |

The root domain **Infra.Shared.Customer** is an empty root domain and contains only the DNS servers to internet and the Certificate authority.

**Systems. Infra.Shared.Customer** is a child domain of Infra.Shared.Customer and contains the main resources like DNS, DHCP, IIS, Citrix of etc. of the child domains EBM.Infra.Shared.Customer. This domain is centralized and is only at Customer however several

companies are using different services of the systems.infra.shared.Customer like Citrix, File Server, Print Server and DNS

**EBM.Infra.Shared.Customer** is a child domain of Infra.Shared.Customer and is on the same level as the Systems domain. This domain contains all the servers and the information of all the users and computers resources. This domain is used worldwide.

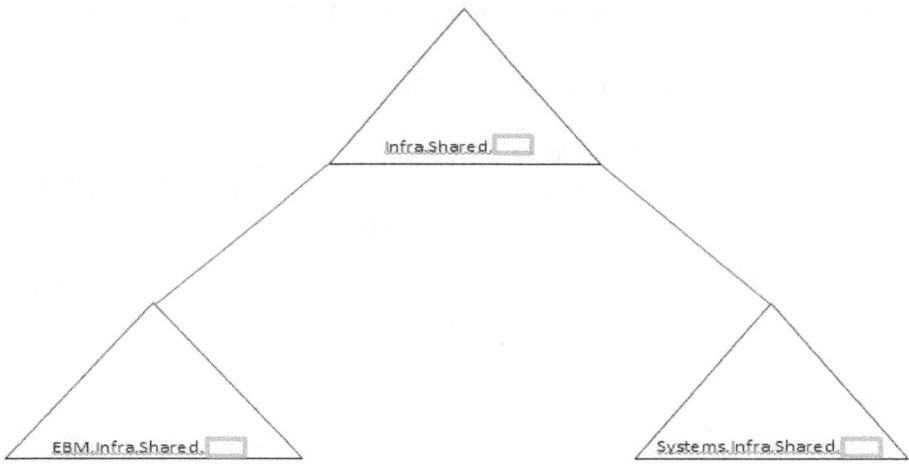

### 2.4.12  Different Services

#### 2.4.12.1      Domain Controllers

The domain controllers of Infra.shared.Customer and systems.infra.shared.Customer are only located in Belgium on the main and disaster recovery site.
The EBM domain controllers are across all the companies of Customer that are already migrated to the shared Customer domain. The ones in the main site CUS are mainly used to logon onto Citrix.
On sites with more than 40 users a domain controller is installed however only 1 full domain controller will be installed per company preferable where the IT staff is located. If the same company has multiple sites with more than 40 users a Read Only Domain Controller (RODC) will be installed on these sites. Exceptions are possible

#### 2.4.12.2      File Servers

Companies using full Citrix have their data located at Customer on the main site CUS. This data is replicated to the disaster recovery site every 2 hours. Local file servers are still available to store data of applications that are not approved to run in Citrix. Ex. AutoCad,...
All file servers are members of the Systems domain except when a company/site is not using full Citrix. In this case the company, user mail data and all other services will stay locally. The servers are will be installed in the EBM domain.
The folder structure is explained in topic Data structure in Customer.

### 2.4.12.3    DNS Server

DNS service is AD integrated and forest wide. All client computers are pointing to the central DNS servers in Customer except clients not on Full Citrix. Whenever a company is not on Citrix or is only using only SAP in Citrix these users will point to the DNS service located on the EBM domain controller on their site if it exist. Only DNS services on site (not using full Citrix) can access or are allowed to use root hints. The DNS service located in Customer or on the Disaster recovery site are not allowed to go to the internet and therefore the root hints must be deleted. Specific applications or servers that need to resolve DNS queries to the internet uses the DNS servers in the infra.shared.Customer.

All DNS zones, whatever the domain they serve, are replicated to all domain controllers through the entire forest (INFRA, SYSTEMS, EBM). Each and every DNS zone exists on every domain controller, whether or not DNS has been installed.

### 2.4.12.4    McAfee

On every client or server a MacAfee client is installed. Only clients in the Shared Customer forest will get connected to the EPO server.
Each site with a domain controller will host an EPO repository.
The MacAfee client on notebooks will get updated, when they are in the office by the nearest repository of that site or when they are on the public internet by the EPO server in our DMZ zone.

### 2.4.12.5    WSUS

The WSUS infrastructure is using the same structure as EPO. Each site with a domain controller will host a WSUS repository. The only difference is that if needed also a repository can be built on a site with no domain controller.
Updating Windows will be done in the office by the nearest repository and on the public internet by the WSUS server managed by Customer.

### 2.4.12.6    DHCP

Customer is using 1 central DHCP server to add the scopes of every site/companies. IPhelper are used to delegate the correct IP scopes to the correct sites.

### 2.4.12.7    OU Structure

**EBM.Infra.Shared.Customer**
In the chapter we will explain how the structure of EBM (short name of EBM.Infra.Shared.Customer) in Active Directory is setup.
Besides the Domain Controller Organizational Unit there are 4 other Organizational Units (OU) that are important for every administrator, local IT or CIT.
These OU's are explained in the following section:

**The EBM OU**
This organizational Unit is divided in different sub OU's. Each sub OU reflects a company and can only be managed by the local or shared IT of that company. Helpdesk Customer and CUS are able to do changes on all company's OU.
The EBM OU is only available in the EBM.Infra.Shared.Customer

```
□ 🗐 EBM
   ⊞ 🗐 BEMA
   ⊞ 🗐 BRSI
   ⊞ 🗐 FRSI
      🗐 ARDU
      🗐 ARET
      🗐 UKMA
```

The company OU's are again divided in different sub OU's and contains all the user, computers and security group objects. Those different OU's are:

```
□ 🗐 EBM
   □ 🗐 BEMA
      ⊞ 🗐 Admin
      ⊞ 🗐 Groups
      ⊞ 🗐 Printers
      ⊞ 🗐 Users
      ⊞ 🗐 Workstations
```

### Admin

Contains the security groups that makes It possible to manage the Company's OU, the company's folder structure and the company's computers. Therefore it can't be managed by local IT's. Every change in this OU need to be asked to Helpdesk Customer. It contains the groups:

> GG00MSFILE-NORM004200xP: Managing the companies folder structure
> GG00HLPDSK-NORM004200xP: Managing the company OU in Active Directory
> GG00LOCADM-NORM004200xP: Give local admin rights on all the computers object in the OU workstation of that specific company

### Groups

This OU contains all the security groups for accessing the companies' folders. The local IT can define their own structure in here. Also they can create extra security groups to make it easier to manage their company but the naming convention should always be followed.

### Printers

This OU is mainly used for administrative purposes. Local IT can create printers object to have a better view which printers they have in their company.

### Users
All user objects for that specific company should be created in this OU. The local IT can create extra sub OU's to match their company's structure. Ex. Each site their own OU.

### Workstations
All computer objects need to be created in this OU. This OU has 3 different sub OU's and depending on the type of the client workstation the object should be moved in the correct sub OU. The 3 sub OU's are:

> Desktops
>
> Laptops
>
> Thinclient

The local IT is entitled to create sub OU's below the desktop, laptop and thinclient structure.

## GRP OU
This OU contains different sub OU's to make the management of the Wintel infrastructure easier. This OU cannot be maintained by the local IT. The sub OU's are GG, GG APPL, GG Restricted. DL, GR, GF. They contain the specific roles and functions

```
▲ 📁 GRP
   ▷ 📁 DL
   ▷ 📁 DL RESTRICTED
   ▷ 📁 DT
   ▷ 📁 GF
   ▷ 📁 GG
   ▷ 📁 GG APPL
   ▷ 📁 GG PROJECT
   ▷ 📁 GG PROXY
   ▷ 📁 GG RESTRICTED
   ▷ 📁 GR
   ▷ 📁 GT
   ▷ 📁 Temp
```

### OU DL
This OU contains mostly the domain local groups of different folders on for local file server (on site). Only GG groups can be member of the DL groups. Standard the GG groups in **EBM/%company%/Groups** are member of the DL groups

### OU DL Restricted
This OU contains domain local group which specific access rights. This OU is maintained by the Enterprise Admin. Only GG groups can be a member

### OU DT
Contains the Domain Local Groups which are linked with a technical right. Ex. Local Server Admin. Only GT groups can be a member

### OU GG, GG Project, GG Proxy
Contains the different global groups not linked to a specific company folder. ex. Project drive access (GG Project), specific rights on the proxy (GG Proxy). Only users or other GG groups can be member of these security groups

### OU GG APP

This OU is the only one that can be managed by the local IT. It contains all the groups that give applications to the user through Citrix. Only users or other GG groups can be member of these security groups

### GG OU Restricted
This OU is also restricted to the Enterprise Admin. It contains all the groups with specific rights. Only users or other GG groups can be member of these security groups

### OU GR
Is used for specific roles such as WSUS-, SQL-, file admin. A different Function (GT) can create roles. Roles are used for the ease of the system. Only a user can be a member of the GR group and only a Global group can be "**member off**" (GT, GG, GF)

### OU GF
GF groups are used to create specific functions on the domain. They contain different GT, groups. Different GT groups combined make 1 GF group. Ex. A Local IT needs local admin rights on all the servers of his/her company. The GT groups of those servers with local admin rights will be a "member of" the GF group.
Only users can be **member** of a GF security group.

### OU GT
GT security groups are defined as a technical right. Ex. Local Admin, Print management,…

## SRV OU

```
⊿ SRV
   ▷ APP
   ▷ CITRIX
   ▷ DEPLOY
   ▷ FILE
   ▷ INFRA
   ▷ MAIL
   ▷ PRINT
   ▷ SQL
   ▷ TECH
   ▷ WDSStaging
```

### OU WDSStaging
This OU is only used for staging servers. New server installations with WDS will be put in this OU. There are no GPO applied to this OU. Only RDP is enabled. As soon as the server is ready, it needs to be moved to the correct OU

### OU all others
The other sub OU's in SRV OU contains servers that matches the specific services. Most of these OU's are empty because resources are added in the Systems.Infra.Shared.Customer domain

### Systems.Infra.Shared.Customer
The OU structure of Systems has the similar organization as EBM.Infra.Shared.Customer. Only the OU EBM is not available in systems. This is because there are no different companies on this domain. Systems.Infra.Shared.Customer contains most of the infrastructure servers like the Citrix, DHCP, DNS, WSUS, EPO server….

### Infra.Shared.Customer

Is an empty root domain but has the same organization structure as Systems.Infra.Shared.Customer. The only servers are the DNS servers with external access and Certificate authority servers.

**Active Directory Replication**
Except from three site links properties, all Active Directory Replication related services and settings are on their default values.

In opposite to intra site connection objects, inter site connections are replicating at minimum interval of 15 minutes. In some cases, replication between some main sites should happen as frequent as intra site replication. This can be achieved by enabling change notification on the site link. For more information about change notifications, please refer to:
http://technet.microsoft.com/en-us/library/cc961787.aspx
Following site links have change notification enabled:

- BE01-CUS-BE02-Brussels
- BE01-CUS-BE03-CUS
- BE01-CUS-BE100-DRPSite

### 2.4.12.8 Security groups

The groups in the Customer forest follow a naming convention which tells us what they are used for.

GG 00 7900 – NORM 000900 x CP
GF 00 LOCADM - SPTW0001 x P
GG 00 CITRIX – NORM 000100 x P

They are built as follow:
1. The first two letters define the scope of the group:

**DL (Domain Local)**
Used on the folder structure

**GG (Global group)**
For folder access: used to bind a user with the proper access rights on a folder.
User ➔ GG ➔ DL ➔ Folder

**GF (Function group)**
These groups are used to define a specific function in the environment. Ex. SQL admin, DHCP admin, SSO access

**GR (Role group)**
These groups are used to combine different GF into a role. Ex. Function SQL admin + DHCP admin ➔ Global Admin

**DT (Domain technical)**
These groups are used to combine different GT groups (ex. from different domains) to access technical parts of the environment.

**GT (Global technical)**
These groups are used for users to give access to certain technical parts of the environment. Ex. Local administrator rights on a specific server. Know that different GT groups can be combined with GF groups to create a GR group.

2.  The next two 0's are for filling the gap between the next parts

    **00**

3.  The next 5 or 6 characters describe the application, function or the location in the file and folder structure

    | | |
    |---|---|
    | **CITRIX** | Citrix app related |
    | **7900** | Customer related group |
    | **ESX** | VMware related group |

4.  The hyphen that now comes is the relation between the first and second block in the group name.

5.  The second block's usage depends on the scope of the group. The scope defines the application, folder,

    **NORM**
    Normal group to address standard functions, access rights for folders and applications

    **CONN**
    Only used on folder access. Is used to connect shared data across companies (G drive)

    **Servername**
    Specific function on that specific server

    **Function**
    Mostly used with application groups. Specifies the function/ access rights of an application (ex. ADMIN, VIEWER, HELPDESK)

    In case the security group is used to give folder access to a user read the comment below. In case the group is used for an applications or a function skip 6 and go to 7.

6.  The next 6 characters define the rest of the scope of the group. 000900 will indicate the IT folder in our new file structure.000ADM refers to an administrative role. The different digits can describe the different levels of permissions that can be granted. At this moment that is only 2 levels. Those digits can be used as follow to address a folder on the second level of security:

    Ex. 2$^{nd}$ level (02 Test) in IT root folder on F company drive.

Se_007900_0900_IT\02 Test
GG007900-NORM000**902**xCP (old way)
GG007900-NORM00900x**002**xCP (New way)

7. To make the group more readable an X has been added after the six digits for folder access groups or after the function or server name of the rule number 5

X

For the new naming convention of the security groups used for folder access this first X is used to split the root folder's number ( 900 in our previous example) and the subfolder's number. The subfolder's number exists always out of 3 digits. This subfolder number is always followed by a second X.

8. To finalize the name of the security groups used for folder access, CP or RP is added. C is used for Modify access and R is used for Read Access. The P indicated that the group is used on the Production environment. T or Q can also be used as Test and Quality. Groups used for Applications and functions end with a P/T/Q. there is no need to add C or R.

**Examples**

Folder Access groups
GG007900-NORM000901xCP (old)          or GG007900-NORM000900x001CP (new)

> => Modify rights on the folder 2 on the second level of security in the root folder 900 which is our IT folder.

A function or role group is a bit different in the way that they describe a role or function and no need exists for connectivity or other differentiation.
GF00CUSTOMERFULLADMIN

> ➔ full administration function, this means all administrator roles.

GR00WSUS-FULLADMINxP

> ➔ WSUS administrator

GG006000-NORMLOCADMxP

> ➔ Local admin on the 6000 company computers.

GT00LOCADM-SPTW0001xT

> ➔ Local Admin on SPTW0001xT

**Comment about group nesting**
Not all groups can be added into the others, that's why we use the default schema of Microsoft.
You place a domain local (dl) group on a folder and grant this group the proper permissions.
Then we add a global group (GG) into the DL.
Users can then be added into the GG or other GG can be added in there.

### 2.4.13 Information Library

| Context | File name | Content |
| --- | --- | --- |
| **Detailed VM cluster info** | **CUS - Virtual Environment** | Technical info from VM |

| | All Info- VM - Hosts – DS.xlsx | environment |
|---|---|---|
| **Application list** | **Virtual Server List of Applications.pdf** | Information on the applications hosted in VM environment |
| **Server naming convention** | **Server Design Documents - Naming Convention Guideline.pdf** | |
| **SQL Server Inventory** | **SQL Server Inventory.xlsx** | |
| **AD Inventory** | **AD Server Inventory.xlsx** | |

## 2.5 LOT 4 – Citrix

Citrix is used as primary application delivery concept for all central applications. The production Citrix farm is hosted in the datacenter in CUS and we have a disaster recovery Citrix farm in the datacenter in Colocation.

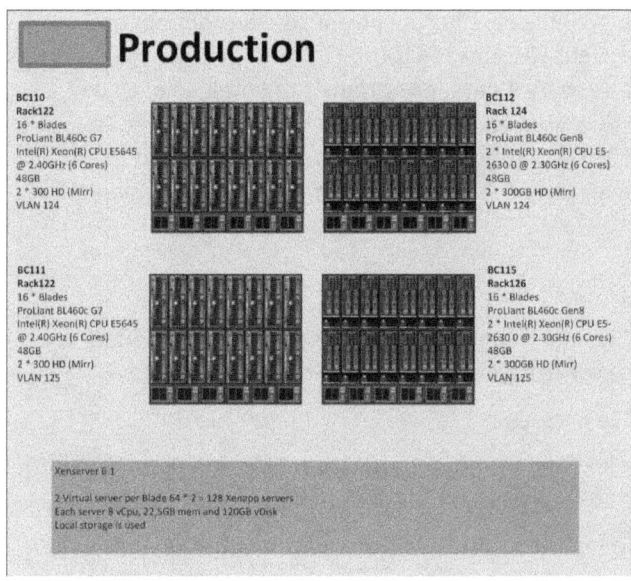

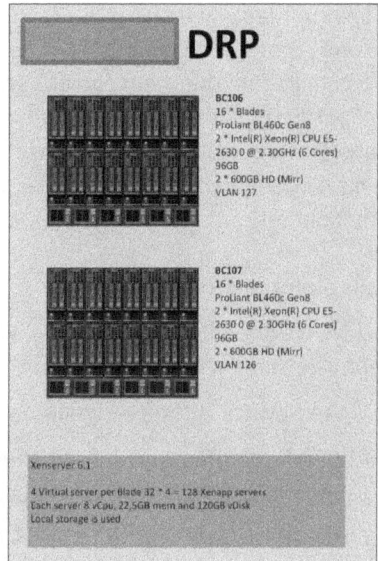

### 2.5.1 Functionality

The XenApp environment is used to access all applications and data made available in CUS. A list of all offered applications can be found in file **Published applications in Citrix.xlsx**. The most important and most used applications are listed here.

- SAP GUI : ERP
- Bex Analyzer
- Max Web UI : CRM
- Lotus Notes 8.5.2 : Used for access mail and databases
- Excel 2010
- Word 2010
- PowerPoint 2010
- Access 2010
- Cognos: Financial reporting
- Certo
- Onguard
- Project 2010
- Visio 2010
- FMS
- MIS

- …

All applications are offered by all 128 servers. All users can be serviced by all servers.

### 2.5.2 Availability

The environment is available 24/7. There are at least 100 concurrent sessions during the weekend night, the max amount of concurrent sessions is 4100.
All servers go into maintenance in 3 groups every 3 days, one group is maintained each day. 8 hours before maintenance is started, no new user sessions are accepted on the server. 15 minutes before maintenance is started, the system sends messages to the connected users. When maintenance is initiated, users are logged off. These users can open a new session on another server.

**Details maintenance schedule**

Every 3 days one third of the servers will go into maintenance. The maintenance schedule on acceptance servers is the same. But these servers will go into maintenance every 2 days.

This is the flow for the server maintenance program:

7:00 PM: Maintenance starts. Server load index will go to full load. Meaning users on the server will work as they are used to. The server stops accepting new connections.

7:10 PM: McAfee scan is initiated. No impact for the users on the server.

Between 3 AM and 3:30 AM: The user starts receiving messages that the server will restart, asking the user to save their work and close the session. They can immediately start a new session on a server that is not in maintenance. The message occurs at 15,10,5,3,2,1 minutes before forced log-off.

Between 3:15 AM and 3:45 AM: Users are logged off from the server

Between 3:20 AM and 4:00 AM: Server restart

Between 3:20 AM and 4:20 AM: Maintenance tasks are ran

Between 3:20 AM and 4:20 AM: Changes are implemented

Between 4:00 AM and 4:30 AM: Server restart

Between 4:00 AM and 4:30 AM: Server maintenance ended. New user logon is accepted.

### 2.5.3 Business Criticality

All ERP, File, Mail and Applications for 7000 users are accessed via the Citrix XenApp environment. In case of failure, a Disaster recovery farm will service all users. The disaster recovery farm is hosted in the datacenter Colocation. 2 bladecenters are reserved to host the 128 XenApp servers. These will handle all user connections and offer all applications when in disaster recovery mode. In normal production mode these servers are not used. DRP XenApp servers are kept up to date in the same way as production servers.

### 2.5.4 Business Purpose

The business purpose is dependent on the type of users. There are 2 types of users. Full Citrix users and SAP Citrix users:

- Full Citrix users: Company data has been centralized in the CUS datacenter. Most of the company data is accessed via the Citrix client. The complete office suite is used together with Lotus Notes for email. SAP GUI is used for access to the ERP data. Published applications and desktops are used with ratio of 90/10
- SAP Citrix users: These users only make use of the SAP applications on Citrix.

Access to the ERP package, company data and mail is provided via the Citrix environment. This makes the XenApp environment strategically very important.

### 2.5.5    DTAP Concept

Within the Citrix environment we have setup a non-production in which the infrastructure team can test changes, updates, new applications, … . If this test goes ok, we apply the same in the Acceptance environment which is used by a limited set of business users. After some time (couple of weeks depending of the change), this goes into the production environment using the maintenance schedules as described above.

### 2.5.6    Citrix System info

Detailed system info can be found in file **Citrix system info.docx**

### 2.5.7    Information Library

| Context | File name | Content |
|---|---|---|
| **Detailed System info** | **Citrix system info.pdf** | |
| **Citrix Hardware** | **Citrix hardware.xlsx and Citrix.vsd** | |
| **Applications** | **Published applications in Citrix.xlsx** | List of all applications published in Citrix |

### 2.5.8    DTAP in the XenApp environment

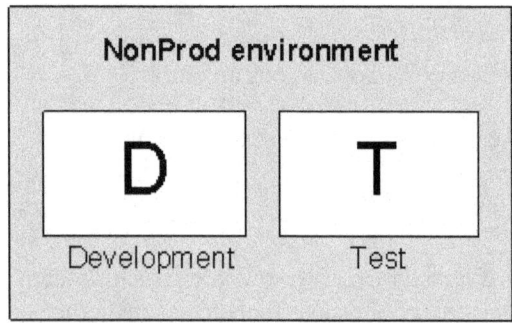

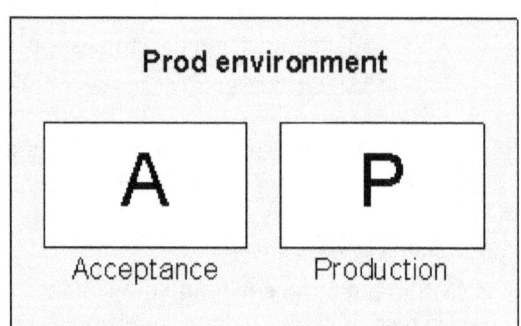

As you can see there are 2 physical environments: the non-production and the production environment. These 2 environments are setup as independent as possible. This separation is key to make sure that tests do not interfere with production.

When new changes are required like patches, programs, configuration changes, ... the change will pass 4 phases: Development, Test, Acceptance, Production. The development and testing is done in the NonProd environment, the acceptance and production phase will take place in the production environment.

**NonProd environment**
The non-production environment uses its own resources that are an exact copy of the production environment. This way we should get a good understanding of what impact changes will have on the production environment when implemented.
As mentioned before the Non production environment will take care of 2 change phases. For the Citrix resources this means that at least 2 servers per silo are available. One used for development of changes, the other used for testing purposes. For other resources like AD servers, file servers,... the same servers are be used for development and testing of changes.

### Development
In the development phase the change will be analyzed and automated or a manual procedure is created. The automation part is the creation of a command line or script that will implement the change without manual actions. For the Citrix resources it is very important to use only automated tasks.

### Test
When the change is automated it will be deployed to all targeted servers. In case of a procedure the manual action is done on all targeted servers.

In most cases only technical tests are done in the test phase. These tests will be performed by the IT team and can include more detailed tests by first line IT staff. Technical tests only ensure that the application is working not the functionality of the program.

When the changes can have effect on the operation of the environment a battery of standard tests will be performed. These tests cover basic functionality of the environment for example the ability to open a Citrix session and edit a word document. A document with these standard tests should be created; this document will list all tests and a procedure how to perform the test. This should not be a static document and as required test can be added or removed.

### Prod Environment

In order to accept changes some servers from the production environment will be used as dedicated acceptance servers.

#### Acceptance

Acceptance servers are equal to production servers in any way. The only difference is that changes are deployed to these servers first. Users assigned to work on acceptance servers are acceptance users. They do their day to day tasks on these servers as all others users. The users allowed in acceptance are knowledge workers, they should have a good understanding of the fact that they are using acceptance resources to do their job and if he/she has problems these are reported to the helpdesk with a note that they are working in acceptance. These users are a mix of FLIT and normal users.

What if a change makes the acceptance environment unusable for users to do their tasks, in that case the acceptance user will switch to the production servers and continue to work on these servers until the acceptance servers are up and running again.

#### Production

Production is the big mass of available servers. Changes are only deployed to these servers if they do not cause problems on the acceptance servers. Changes are deployed to the servers during the maintenance window of the server. This is every 3 days. This means that it takes 3 days before a change is fully deployed in the production environment.

### Changes

All changes are assigned a change ID. The progress of the installation is followed up and documented as seen in the example below.

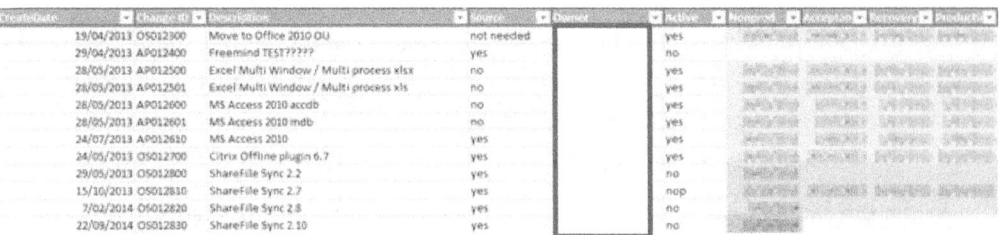

| CreateDate | ChangeID | Description | Source | Owner | Active | Nonprod | Acceptance | Recovery | Production |
|---|---|---|---|---|---|---|---|---|---|
| 19/04/2013 | OS012300 | Move to Office 2010 OU | not needed | | yes | | | | |
| 29/04/2013 | AP012400 | Freemind TEST?????? | yes | | no | | | | |
| 28/05/2013 | AP012500 | Excel Multi Window / Multi process xlsx | no | | yes | | | | |
| 28/05/2013 | AP012501 | Excel Multi Window / Multi process xls | no | | yes | | | | |
| 28/05/2013 | AP012600 | MS Access 2010 accdb | no | | yes | | | | |
| 28/05/2013 | AP012601 | MS Access 2010 mdb | no | | yes | | | | |
| 24/07/2013 | AP012610 | MS Access 2010 | yes | | yes | | | | |
| 24/05/2013 | OS012700 | Citrix Offline plugin 6.7 | yes | | yes | | | | |
| 29/05/2013 | OS012800 | ShareFile Sync 2.2 | yes | | no | | | | |
| 15/10/2013 | OS012810 | ShareFile Sync 2.7 | yes | | nop | | | | |
| 7/02/2014 | OS012820 | ShareFile Sync 2.8 | yes | | no | | | | |
| 22/09/2014 | OS012830 | ShareFile Sync 2.10 | yes | | no | | | | |

### 2.5.9 Number of users

There are +- 7000 users making use of applications on the Citrix environment in 3 months. Most sessions on a concurrent base is 4100.

### 2.5.10 Monitoring

All servers are monitored and by multiple tools.

**Control up**

Monitors all important servers' performance counters on a real time base.

These counters give an overall impression on the performance and user perception.

The sum of these counters provides a stress level for each server. Once the server reaches a certain stress level immediate action is taken to solve the problem. This action can be taken from within the tool.

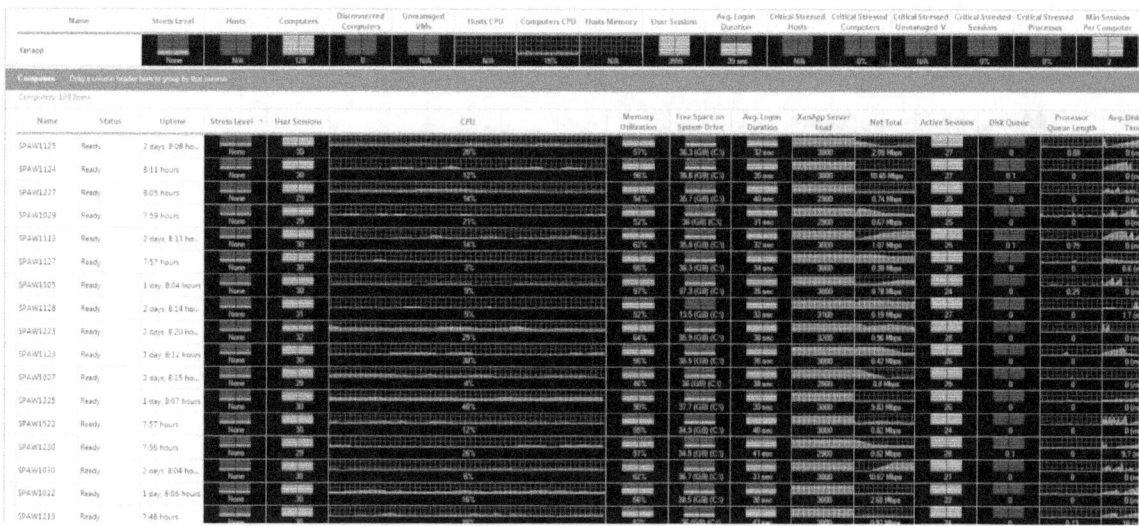

Next to this certain general counters are considered important to have an overall look on the performance of the environment. These are:

- Total number of sessions
- Avg stress level
- Avg logon duration
- Avg number of users
- Avg CPU usage
- Avg memory usage
- Max Nbr if sessions per computer

### Historical data Edge Sight

All historical data is monitored and made available by Citrix Edge sight. This tool provides reports on historical application usage and user performance.

### 2.5.11  InFlight projects

Currently we are making a new design to upgrade the Citrix environment to version 7.6 and integrate NetScaler as well as gateway. This project will start in 2015 and will continue in 20xx.

### 2.5.12 Naming convention

#### 2.5.12.1       XenApp server

**SPAWaabb (Production)**
- SPAW : Server Production Application Windows
- aa : Bladecenter number. Can be 10,11,12,16
- bb : Sequence number virtual machine. From 1 to 32

**SRAWaabb (DR)**
- SPAW : Server Recovery Application Windows
- aa : Bladecenter number. Can be 06,07
- bb : Sequence number virtual machine. From 1 to 64

#### 2.5.12.2       XenServer

**SPVXaaaa (Production)**
- SPVX : Server Production Virtual Linux
- aaaa : Sequence number from 1 to 9999

**SRVXaaaa (Production)**
- SRVX : Server Recovery Virtual Linux
- aaaa : Sequence number from 1 to 9999

## 2.6 LOT 5 – MAIL & STORAGE

### 2.6.1 Mail & Groupware

Lotus Notes is used as Customer Group email and collaboration platform which is centrally managed by Customer. Within the Datacenters of Customer, the central Customer Group infrastructure is hosted which is in scope of this RFP. On top, Customer manages Lotus Notes servers outside of Europe for which the client cannot be centralized yet via Citrix due to the high latency.

The Lotus notes Email infrastructure consists of an Email cluster and storage on NetApp in the datacenter in CUS. The backups are done using NetApp technology and mirrored to the datacenter in Colocation. In the DRP datacenter, we mirror the archive server and both mail clusters.

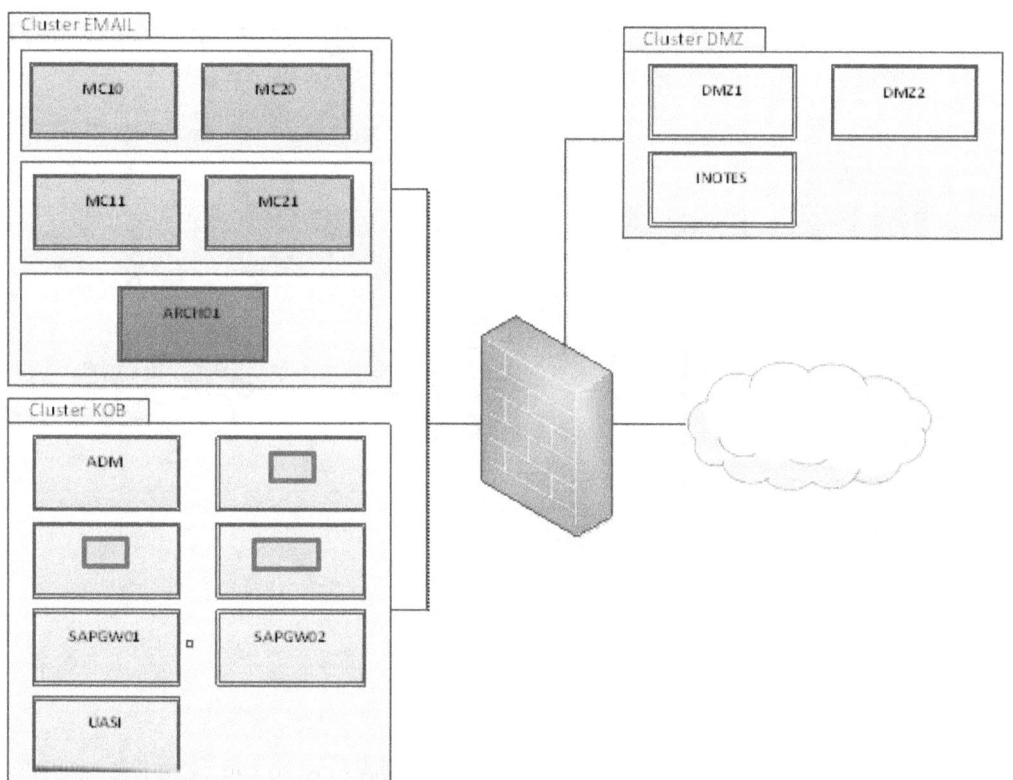

Within Customer Group, several Lotus notes email domains exist. We have already consolidated a lot during centralization projects but still some are left. Only the domains Customer Group and BEDMZ00 are located at CUS.

SapGroupware01

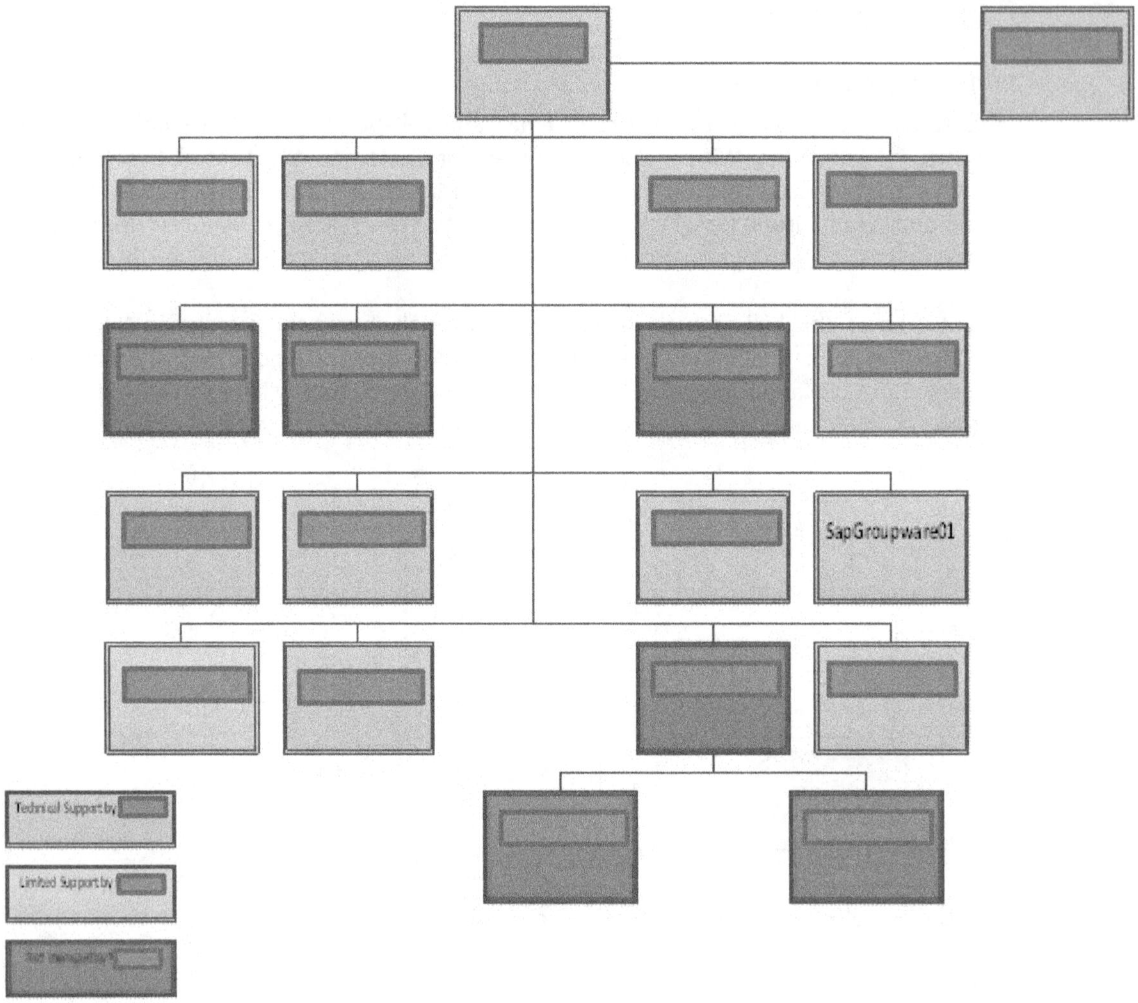

Technical Support by

Limited Support by

Not managed by

**Backup retentions**

The following table is an overview of the snapshots taken by SnapCreator. Cron gives the time of day the snapshots are taken, while Count is the number of days that the snapshots are kept. Snapshots older than Count are deleted.

| Profile | Config | Retention | Count | Cron |
|---|---|---|---|---|
| **spmw0001** | | | | |
| | LotusNotes_Systemdrives | hourly | 48 | |
| | LotusNotes_data | hourly | 30 | 6h00 & 20h00 |
| | CUS_data | hourly | 90 | |
| | | daily | 7 | |
| | | weekly | 4 | |
| | test_ps | hourly | 1 | |
| **spmw0003** | | | | |
| | CUS3_data | hourly | 90 | 0h40, 7h40,12h40,16h40,20h40 |
| | | daily | 7 | |
| | | weekly | 4 | |
| **spmw0008** | | | | |
| | spmw0008_online | hourly | 30 | 7h00,13h00,22h00 |
| **spmw0009** | | | | |
| | spmw0009_online | hourly | 30 | 7h30,13h30,22h30 |
| **spmw0010** | | | | |
| | spmw0010_online | hourly | 30 | 22h15 |
| **spmw0011** | | | | |
| | spmw0011_online | hourly | 30 | 22h45 |
| **spmw0025** | | | | |
| | be_notes_arch01 | daily | 30 | 20h00 |

For all the domino servers, also a snapvault backup copy is created at Colocation, once a day. The retention is 84 days (12 weeks).

### 2.6.2 InFlight projects

We are currently running an archive project to reduce the size of the active mailbox.
We are currently investigating the cloud solution from IBM for Lotus Notes (IBM SmartCloud).

### 2.6.3 Lotus Notes client

Most user Lotus Notes accounts are defined in the Customer Group domain, i.e. +7000 users.
Most users within this Customer Group domain, own a mailbox on the mail infrastructure in the CUS Datacenter (cluster MC10/MC11 (+6000 users)).

When the network bandwidth from a certain Customer company is sufficient (MPLS), then mailboxes for the end-users are centralized on the MC10/11. In this case, we provide a "Citrix/Lotus Notes" client. Both the Notes client and the mailbox are located at the mail infrastructure in the datacenter in CUS. The Lotus Notes application we provide is the basic client 8.5.2 FP3. There are plans to deploy release 9 client in the near future, but no roll-out plan or schedule is determined yet.

We always strongly encourage "Citrix/Lotus Notes" clients for obvious reasons (bandwidth consumption, centralized management of the clients, ...), but some exceptions are allowed for certain groups. Typically, end-users who are rarely working from an Customer office (e.g. sales reps) can make use of local Lotus Notes clients. On such cases, a local replica could be found on their desktop. Installation and configuration are the responsibility of the IT within the companies.

Some companies outside of Europe (geographically too remote from Belgium), are hosting a separate Lotus Domino server. End-users contact these servers by a local Lotus Notes client. The hardware and support is done by the IT within the company and we act as Third Line support.

### 2.6.4    Web Browser

Besides the Notes client, end-users do have the possibility to use a browser in order to access their mailbox. This method is referred as Web Mail or iNotes. Nevertheless, even when this technology is available and fully operational, this access method is not yet promoted by Customer.

### 2.6.5    Remote Access

Several options are possible for end-users.

**User VPN-Connection**
Laptops with VPN-connection make it possible to access Citrix applications and hence also the possibility to launch the "Citrix/Lotus Notes" application. By doing this, end-users have the same Notes environment (workspace, mailbox) as they are entitled to in the local office.

**Tablets with Lotus Traveler**
Recent mail messages are available on the tablet (also off-line) and mail communication is possible on the mail app. Mind that end-users have limited possibilities when it comes to attachments.

**Tablets with Lotus Traveler and Citrix Receiver**
Recent mail messages are available on the tablet (also off-line) and mail communication is possible on the mail app. Mind that end-users have limited possibilities when it comes to attachments.
By using the Citrix receiver, tablet users can also launch Citrix applications, including the "Citrix/Lotus Notes" application.

**Local Lotus Notes client**
When using off-line, end-users can operate on a local replica of their mailbox. This option can be considered for users that travel frequently and they need to access their local mailbox anytime.

By using a secured VPN connection, the local replica can be synchronized with the mailbox on the server. As mentioned earlier, the end-user can then also launch Citrix applications.

**BlackBerry**

A BlackBerry Enterprise Server at Customer was installed in order to deliver mail connectivity to BlackBerry devices. Customer does not promote new deployments and expects that the number of users will drop in the future.

### 2.6.6    Email Archiving

The archiving solution is currently being deployed for all mailboxes on the mail cluster MC10/11. Besides the fact that it will result in faster mailbox operations, it provides also some improvements, such as an enhanced search function and an automated archiving mechanism by means of Lotus Notes policies.

The IONet external archiving solution is configured and this is active on BE_NOTES_ARCH01. We expect that all mailboxes will have an active archive counterpart by the end of the October 2015.

We estimate that, by the end of the project, the total storage of the archiving will be 20 TB.

Some characteristics:
- All mail messages older than 6 months are automatically archived.
- The archive has the same folder-structure as your mailbox.
- Calendar Items are NOT archived
- E-mails marked with a "follow-up" flag are not archived!
- The archive mailbox can be opened from within the operational mailbox (archiving+open) or by clicking its mailbox icon on the Lotus Notes workspace.

Some companies (Ivarsson Denmark, some UK offices, ...) require that full archiving will be done when their users leave the company. In this case, the archiving mailbox will not be deleted.

### 2.6.7    Anti-spam / anti-virus

**Overview**

The Customer Group mail infrastructure makes use of the Barracuda anti-spam and anti-virus appliance to safeguard a healthy mail environment. Those devices are hosted in the datacenter in CUS with an active standby in Colocation.

Below, you will find some characteristics about its configuration.

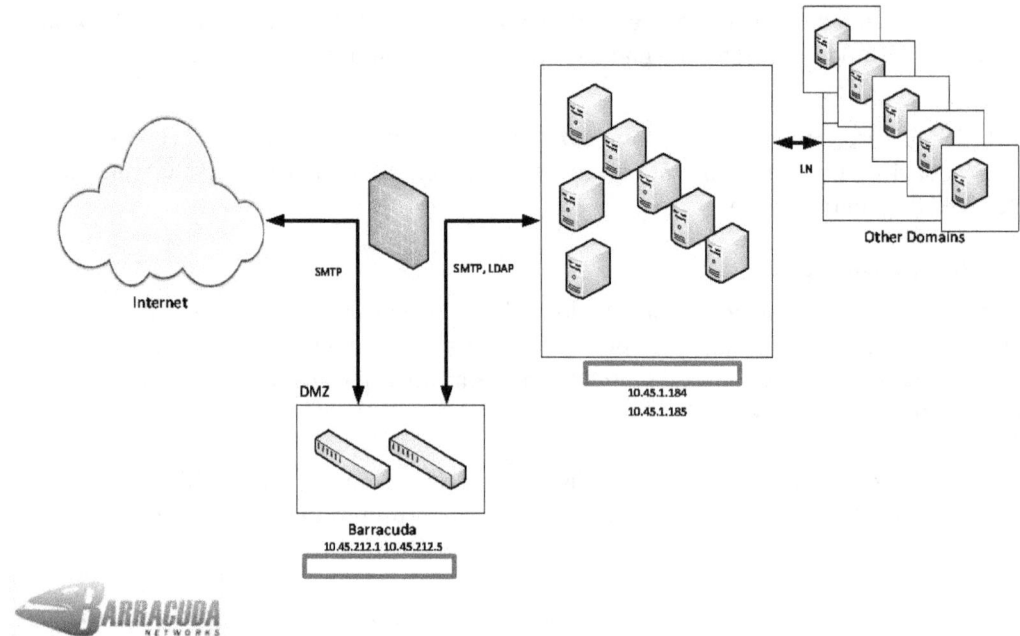

## Affected Lotus Notes domains
Not all Lotus Notes domains are making use of Customer's Barracuda solution. In that case, incoming and outgoing mail are handled by a separate mail gateway and as a consequence not handled or supported by Customer.

Included domains
- Customer Group
- Creaton
- IETE00
- Customer-OpCo (OpCo Italy)
- IDET00

Excluded domains
- ARET00
- COSL00
- CLPZ00
- CLET00
- DEPR00
- OpCo-Ukraine
- PEET01

## Incoming mails
Three levels can be identified:
Allowed (spam-score is lower than 3)
Quarantined (spam-score between 3 and 6)
Blocked (spam-score higher than 6)

We're not tagging suspected mail messages. These are added to the quarantine digest which is sent to end-users on a daily basis.

LDAP-verification of the existence of an e-mail address (LDAP-server = BE_NOTES_ADM)
    Anti-spoofing (external exceptions can be accepted)
    Fingerprint analysis
    Image Analysis
End-users have a personal page where they can fine-tune some default anti-spam settings.

### 2.6.8 Email Usage

We don't maintain statistics about internal mail routing for the moment.
The numbers below reflect the number of mail messages that passed the Barracuda servers in the past months 2015.

|  | May 2015 | June 2015 | July 2015 |
|---|---|---|---|
| **Bad recipients** | 326.273 | 273.847 | 259.949 |
| **Spam** | 841.374 | 750.499 | 978.345 |
| **Viruses** | 36.401 | 26.395 | 28.778 |
| **Quarantined** | 132.726 | 134.999 | 158.248 |
| **Allowed** | 1.036.453 | 1.112.328 | 1.093.945 |
| **Total** | **2.373.227** | **2.320.614** | **2.519.732** |

### 2.6.9 GroupWare

Our Lotus Notes environment is also linked with SAP via a Groupware system. This infrastructure is part of the virtual environment and synchronizes data (visits, …) from our central SAP CRM system into the Lotus Notes email and calendar of a selected group of users (e.g. sales reps). The technical configuration is done by third line system engineers in collaboration with SAP functional consultants.

### 2.6.10 Lotus Notes Applications

We offer a Lotus Notes applications platform that is hosted on our central Virtual environment. Our responsibility is to provide the platform up until Lotus Domino level on which the Customer companies can develop applications.
Currently we have listed over 1000 applications and due to the diversity of applications that is hosted on this platform, the criticality varies.

### 2.6.11 Information Library

| Context | File name | Content |
|---|---|---|
| **Lotus Notes Server overview** | **Lotus Domino server overview.xlsx** | Technical info on Lotus notes servers |

## 2.7 Shared by LOTs – Data Centre Operations & Infrastructure

Please describe the current processes and procedures for the following domains. If specific tools are used please provide detailed information.

### 2.7.1    Monitoring

Please describe:

1.  Several monitoring solutions are in place
    SCOM - Wintel environment to monitor availability and utilization of Wintel environment
    Control Up – To monitor usage and performance of Citrix environment
    SolarWinds virtualization manager – to monitor the availability and utilization of the VMware environment
    Solution Manager - to monitor availability and usage of SAP systems
    Solarwinds Network monitor – To monitor the availability and usage of LAN and WAN
    Lotus Notes monitoring is done with an own developed solution to monitor the availability of an address book on the different lotus notes servers. Alerts are sent to service desk and system engineers who will respond based on criticality.

2.  Server availability is monitored by the different monitoring solutions. The monthly report to provide the availability of the central systems is made manually.

3.  Peak periods of the systems are between 10:00 and 16:00 CET and during period end closing (first 3-5 working days of each month)

4.  Critical messages are sent to service desk so they can inform system engineers in example when a system is down. Other messages are sent to the system engineer teams who handle during business hours.

5.  The service desk has a daily check list which describes the checks that are to be done (example: Daily check of NetApp backups). This check list is maintained in a Lotus Notes database.

### 2.7.2    Backup/Restore and Tape operations

1.  Backup procedures have already been described in the separate sections.

2.  Restores of files can be done by the user and for other systems (like mailbox), this happens only occasionally and can be done service desk or system engineers.

### 2.7.3    Governance

Please describe the current ICT organization, IT management culture and maturity.

Customer is the central IT organization of the Customer Group. The infrastructure team is responsible for all central systems as described above and has no control on the local IT (who manages the local infrastructure (LAN, pc, printers, etc.)) which is in the companies. Local IT is also the SPOC for the end users and they will contact central IT via the service desk in India in case of central incidents or requests.

Local IT is in close collaboration with central IT in order to discuss, plan and execute infrastructure related tasks.

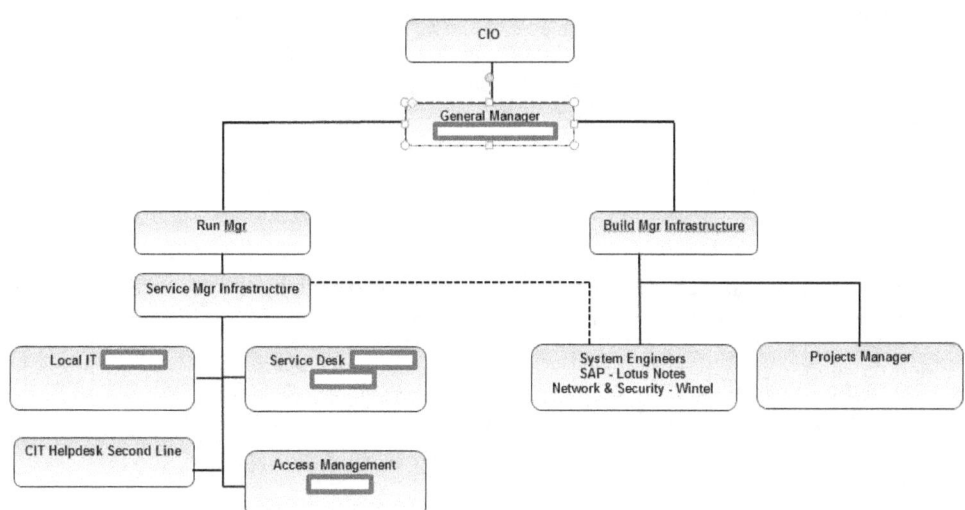

We have a yearly budget cycle in which we look into new systems, growth of systems based on planned centralization projects or other improvements, changes etc…
We are working on a new IT strategy which will have an impact on the future IT landscape. We will update you throughout the RFP phases when more information is available.

Infrastructure projects are managed according to PRINCE2, which has been adopted to the Customer approach. Further description can be found in the PM Guidelines.

### 2.7.4   Process Management and Procedures Manual

We have implemented already some ITIL processes into the Customer infrastructure area and will continue to develop this further. Currently we have 3 processes in place

1. Incident management process

2. Request management process

3. Planned maintenance process

### 2.7.5   Service Level Management and Reporting

Service Level Reporting of the last 6 months in attachment

### 2.7.6   Business Continuity and Disaster Recovery

Disaster Recovery plan is provided as attachment.
Every year, a DR test is executed.
In 2015, we have already done a limited test (scope only AS400 and AIX to test synchronization between SAP systems in a DR scenario) in preparation of the larger test that we plan by end of 2015.

### 2.7.7 Security Management

Customer and the Customer group have different procedures in place, from which some are under review. These procedures describe how Customer and the Customer group want to manage their electronic information and communication systems (infrastructure, systems, applications, etc.), and how these systems need to be protected (keep information confidential and private where needed), assure the integrity of the information and keep the information and systems available.

The Customer IS Security and Risk Policy describes at the highest level how Customer handles the IT systems.

The Customer IS Security and Risk SOP v 1.0 goes a bit more in detail, but remains relatively high level, and gives an overview of how Information systems are managed.

All these documents are for Customer and/or Customer internal use only, however are shared with the service providers as covered and protected by the signed NDA's.

### 2.7.7.1 Information Library

| Context | File name | Content |
|---|---|---|
| **Security Policies** | **150904 - annex - Policy BEMA-ISPOL-00001-001 Customer IS Security and Risk Policy v 1.0.pdf** | Security and Risk Policies |
| | **150904 - annex - Procedures BEMA-ISSOP-00001-001 Customer IS Security and Risk.pdf** | Security and Risk Procedures |
| | **150409 - annex - Data Processing Agreement template.doc** | Template for contracts |
| | **150904 - annex - 2011_Code_of_Business_Conduct.pdf** | Code of Business conduct with some applicable supplier parts for vendor |

### 2.7.8 Asset Management

No Asset management (nor CMDB) solution is in place. An inventory of all infrastructure exists in excel files which is kept up to date by the infrastructure team.

As the infrastructure changes, the situation as described will be changed by the time of the lift and shift phase. We will foresee an update during the RFP process for major changes.

### 2.7.9 Release Management

We do not have an official release management process.

### 2.7.10 Incident and Request Management

LANDesk is used as central ticketing system. Currently we have 2 processes in place: Incident and request management.
Both are fully operational within LANDesk and according to the SLAs with the business.
Reporting is done on a monthly basis using Crystal reports.

A knowledge database is kept up to date for the service desk in Lotus Notes.

### 2.7.11 Performance Management and Capacity Planning

We have different performance metrics that we keep track off.

For the SAP production systems we have provided the Early Watch Reports as first information on performance but we keep track of the performance of specific transactions which have been defined as business critical as well as a good reference point for general performance as they are in the top list of most used transactions.
In the ECC System, we monitor certain transactions in the Sales and Distribution area (VA01, VA02, VL06O, VT02) which are looked at in case of general performance complaints or after major changes (upgrade, hardware change, software changes etc...). In the SAP CRM system, we have a similar set of transactions that are monitored.
System tuning for SAP is done with infrastructure and application teams.

Another key player in performance (mainly for end users) is Citrix. As explained in the Citrix chapter, we use ControlUp for real-time monitoring of the Citrix environment. Highly loaded servers are easily visible and actions are taken if needed to improve performance (e.g.: long logon time for certain users, ...) for end users.

The history of performance numbers we have will be very valuable when we move to the new datacenter as well as to define KPIs and SLAs towards the service partners.

### 2.7.12 Change Management

Small changes can be discussed within the teams and bigger changes are discussed within the Bi-weekly infrastructure meeting. Changes can be requested from different sources.
Upgrades of servers are to be agreed with the business owner.
Execution of the change is to be announced to the business and is in most cases planned during the agreed maintenance windows.

### 2.7.13 Licensing

We share some license information which is closely linked to infrastructure. It is not our intention to share all information nor to transfer licenses. The aim is to look for the best solution in this context when phase 2 will be in place for each of the domains. The attached information gives an overview of the different Software contracts / vendors in scope for the project *Customer Datacenter Outsourcing* Most Software contracts described in this document will require a more detailed discussion and / or content is described in other sections. Further in the RFP process Customer will provide more details if necessary and where needed and depending on the contractual agreements in place with vendors.

### 2.7.13.1 Information Library

| Context | File name | Content |
| --- | --- | --- |
| Licensing overview | **SW_overview_Customer.xlsx** | VMware, SAP, Citrix, Microsoft, IBM |

### 2.7.14 Service Desk – Out-of-scope

The first line service desk is out-of-scope and is managed by ....

### 2.7.15 Workplace – Out-of-scope

Workplaces are managed by the local IT of each company and are out-of-scope.

### 2.7.16 Information Library

| Context | File name | Content |
| --- | --- | --- |
| **Project Management guidelines** | **PM Guidelines infra.pdf** | |
| **Incident Process** | **MQMS -PRC-Incident-Management v0.5.pdf** | |
| **Request Process** | **MQMS -PRC-Request_Fulfilment v0.5.pdf** | |
| **Planned maintenance process** | **QPR ProcessDesigner - 07.06.02.01 IT Planned Maintenance.pdf** | |
| **Business Continuity and Disaster recovery** | **Disaster Recovery Customer CUS 20150901.pdf** | |
| **Monthly Report** | **General Report 2015-01.pdf** | Report on availability, SLAs from central infrastructure |

## 3 REQUEST FOR PROPOSAL – ANSWER TEMPLATES & REQUIREMENTS

The tenderer, vendor, must structure his offer so that all questions and criteria are discussed in the same order and ABC sub numbering of this chapter. Use all provided tables. Vendor may reuse this document (only for answering this RFP):

- Part A. PRICING
- Part B. QUALITY – LIFT & SHIFT
- Part C. QUALITY – MANAGED SERVICES
- Part D. HR TRACK

Provide the following Contact details:

| Contact Details | | |
|---|---|---|
| ID | Question | Answer |
| 1. | Name of organization? | |
| 2. | Contact person? | |
| 3. | Visiting address? | |
| 4. | Postcode & place (visiting address)? | |
| 5. | Postal address? | |
| 6. | Postcode and place (postal address) | |
| 7. | Country | |
| 8. | Telephone number | |
| 9. | Contact person's e-mail address | |
| 10. | Organization's website | |
| 11. | Chamber of commerce registration number | |

### 3.1 PART A. PRICING (via pricing template A)

Expected delivery of pricing and fine tuning during every Vendor activity stage V1-V4 via external Pricing Template A (not to be embedded within vendor's general answer document).

#### 3.1.1 A1. Pricing - Lift and Shift

Divided into one time migration costs and yearly recurring housing & network costs.
Fill in external Template A.

#### 3.1.2 A2. Pricing - Managed Services

Divided into one time migration costs and yearly recurring costs per lot, including the reduction in housing & managed services when the service of a lot is moved from housing to managed service. Fill in external Template A.

#### 3.1.3 A3. Pricing - Calculated Project Assumption

Includes man day pricing of profiles for formula to calculate assumed pricing total for scoring.
Fill in external Template A.

#### 3.1.4 A4. Pricing - Calculated assumption for growth and hosting costs

A list of housing costs, network & security services and IaaS costs, to calculate (unexpected) growth and hosting costs, or if third party managed services is used on one or more Lots, instead

of awarding the managed service to the housing provider.
Fill in external Template A and deliver your complete pricing catalog.

### 3.2 PART B. QUALITY – LIFT & SHIFT (L&S)

Expected delivery of following parts B1-B5 during V1 within the given templates, defined RFP structure and given tables.

Embed Customer's templates in the answer document, not as appendix, but within the requested ABC-answer structure. The only template not to be included within the answer document is the pricing template A.

All answers in this section apply to the lift & shift stage (preceding managed services).

#### 3.2.1    B1. L&S - Description of housing / colocation services

Answer the following questions:

| Housing | |
| --- | --- |
| **ID** | **Question** |
| 1. | Describe your housing /colocation services succeeding the lift & shift of the initial migration project. |
| 2. | What is the procedure of getting access to your datacenters and how do you prevent your personnel and other customers from accessing Customer's hardware? |

**About Connectivity Addresses**:

Customer will ask their WAN providers for price and turnaround time of a new connection from their offices in Location 1 to vendor's datacenter(s).

- Location 1 to datacenter connection
- New datacenters to MPLS (Provider)

| Connectivity Addresses | | |
| --- | --- | --- |
| **ID** | **Question** | **Answer** |
| 3. | What are the addresses where Customer can connect to (L&S stage)? (list at least two datacenters) | |
| 4. | Can the same connections be used for the Managed Service stage(s)? If not, what are the addresses to connect to? | |
| 5. | Is there need for temporary extra bandwidth during the migration period? If yes, please specify connection address, bandwidth need and period. | |

### 3.2.2 B2. L&S - Project Organization

Experience of the Lift & Shift (L&S) team:

| Lift & Shift Team | |
|---|---|
| ID | Question |
| 1. | Deliver a list of the involved L&S team |
| 2. | Describe the L&S experience per team member |
| 3. | What if a key person leaves the vendor/is unavailable? |

| Transport Risk | | |
|---|---|---|
| ID | Question | Answer |
| 4. | Customer expects that the vendor embeds transportation costs into the migration offer.<br>Up to vendor:<br>Is transportation risk insured by vendor and included in the migration offer? | |

### 3.2.3 B3. L&S - Project Methodology, Tools, Technology

Describe your Project Methodology, Tools and Technology used to execute the lift & shift project.

### 3.2.4 B4. L&S - Demonstrated Capabilities & References (template B) (V1)

Embed the template within this paragraph of your answer document.

### 3.2.5 B5. L&S - Risk File (template C1)

Embed the template within this paragraph of your answer document.

### 3.2.6 B6. L&S - Project Plan & Roadmap (V2)

The RFP answer of this paragraph is divided into two vendor activity periods:
- V1: High level estimation of total project turnaround period and costs in part A.
- V2: Project plan and estimated man days per task and total project turnaround period.

## 3.3 PART C. QUALITY – MANAGED SERVICES (MS)

Expected delivery of C1-C8 during V1. All answers of this section apply to the optional managed service stage(s).

### 3.3.1   C1. MS - Organization

Provide your managed service answers to the following questions.

| General Business Information | |
|---|---|
| **ID** | **Question** |
| 1. | Organizational structure, including organization chart of business/parent company? (Deliver company information, shareholders, structure of mother corporation, legal forms, joint ventures, subsidiaries, partnerships or other relevant relations.) |
| 2. | Geographic spread (worldwide)? |
| 3. | Business policy, vision and mission? |
| 4. | Position relative to your competitors? |
| 5. | Office locations/addresses in Belgium? |

| Cost control & trends | | |
|---|---|---|
| **ID** | **Question** | **Answer** |
| 6. | What provisions are set aside by your company to cover any financial disasters? | |
| 7. | What is your operating result before exceptional income and expenditure for the years 2012, 2013 and 2014? | |
| 8. | State your solvency and liquidity ratios, and your profitability ratio in respect of your total capital, debt capital and shareholders' equity. All ratios must be given for the years 2012, 2013 and 2014. Also add your last two annual reports. | |

| General Requirements (Must haves, Should haves, Could haves, Won't haves) | | | |
|---|---|---|---|
| **ID** | **Requirement** | **MoSCoW** | **Answer** |
| 9. | Do you comply with the *"150409 - annex - Data Processing Agreement template"* (external file). | Must | |

| General Requirements (Must haves, Should haves, Could haves, Won't haves) | | | |
|---|---|---|---|
| **ID** | **Requirement** | **MoSCoW** | **Answer** |
| **10.** | Do you comply with *"150904 - annex - Policy BEMA-ISPOL-00001-001 Customer IS Security and Risk Policy v 1.0.pdf"* and *"150904 - Procedures BEMA-ISSOP-00001-001 Customer IS Security and Risk.pdf"* (external files). | Must | |
| **11.** | Do you comply with the supplier parts within the *"150409 - annex - 2011 Code of Business Conduct.pdf"* for Customer employees (external file)? | Must | |

### 3.3.2    C2. MS - SLA, incl. maintenance windows (Pre V1 - V4)

SLA and Continual Service Improvement (CSI)

| SLA Metrics & KPI implementation | |
| --- | --- |
| **ID** | **Answer** |
| 1. | Describe your standard SLA services and metrics as reference here (C2), for your answers per lot (in C3). |
| 2. | Give example KPI's for monthly/quarterly reports, e.g. how often a certain network bandwidth threshold is reached? |
| 3. | How often do you discuss KPI's and improvement plans? |
| 4. | How often do you offer to change KPI's to keep up with maturing services and customer relation? |

Maintenance. Customer is an international company. Users work around the clock.

| Maintenance Window Calendar | |
| --- | --- |
| **ID** | **Answer** |
| 5. | What do you do, when, how many, and how long are your maintenance windows with planned downtime during a year, especially regarding your shared services? |

Contract and exit plan

| Contract Period – Bonus Malus and Exit Plan | |
| --- | --- |
| **ID** | **Answer** |
| 6. | Describe how the destruction of Principals data is recorded and at what time during or after termination of the agreement, which data, in what way and when will be destroyed? |
| 7. | What do you offer for extending a contract period from one year to two, three or more years? |
| 8. | Customer does not want to come into a situation where the vendor pays a bearable maximum penalty and the situation does not improve. <br><br> When SLA's are not matched, Customer prefers a concrete improvement <u>plan within a given time</u> instead of a penalty fee. When an improvement plan is not delivered in time, and <u>when improvement does not occur in time</u>, Customer prefers a significant penalty such as complete monthly fees. <br><br> What timings (2) do you offer for improvement plan and resolution and what do you offer as significant penalty (driver for improvement)? |
| 9. | What flexibilities do you offer as exit plan? What triggers a free exit? |
| 10. | Can Customer stay at vendor's datacenters and keep using vendor's datacenters for housing services, even after exiting managed services, for competitive pricing based on yearly price index and market evolution? |

### 3.3.3 C2 - General Service Level Requirements for HA, DR and Retention

**Availability**

Customer expects High Availability (HA) within one datacenter zone with no single points of failure. A second datacenter must be available to be used as potential active-active solution to prevent the need for site failover after a disaster in one datacenter.

**Continuity**

Customer expects a Disaster Recovery (DR) Plan involving a second datacenter for recovery and archiving within another availability zone.

Example business Availability and Continuity metrics (no requirement for this RFP):

| SLA | "Gold" | "Silver" | "Bronze" |
|---|---|---|---|
| **HA** <br> **(within 1 DC)** <br> **(highest severity level "P1")** | RTO=5min <br> RPO=5min <br> Max. 3.6hr unplanned down time per month (99.5%) <br> No Single Points of Failure on 1 site | RTO=15min <br> RPO=5min <br> Max. 7.2hr unplanned down time per month (99.0%) <br> No Single Points of Failure on 1 site | RTO=15min <br> RPO=5min <br> Max. 14.4hr unplanned down time per month (98.0%) <br> No Single Points of Failure on 1 site |
| **DR** <br> **(site failover)** | RTO=15min <br> RPO=5min | RTO=4hr <br> RPO=5min | RTO=8hr <br> RPO=24hr |
| **DR** <br> **(due to replicated failure; requiring restore)** | RTO=4hr <br> RPO=5min | RTO=8hr <br> RPO=12hr | RTO=12hr <br> RPO=24hr |
| **Retention period for Backups – VM's** | <= 2 weeks | <= 2 weeks | <= 2 weeks |
| **Retention period for Backups – Data, e.g. from file shares** | 10 years. <br> Good enough: <br> Daily backups: 2 weeks <br> Monthly backups: 12 months <br> Yearly backups: 10 years | 10 years. <br> Good enough: <br> Daily backups: 2 weeks <br> Monthly backups: 12 months <br> Yearly backups: 10 years | 10 years. <br> Good enough: <br> Daily backups: 2 weeks <br> Monthly backups: 12 months <br> Yearly backups: 10 years |
| **Support Availability Window (Service Desk) (All systems run 24/7)** | P1: 24/7 <br> P2-P4: European Office hours | P1: 24/7 <br> P2-P4: European Office hours | P1: 24/7 <br> P2-P4: European Office hours |

Customer will keep their existing central end-user Service Desk as point of contact for accessing vendor's support.

The main difference between gold, silver and bronze is the priority for troubleshooting,

recovery and startup after a disaster. There is almost no difference in High Availability, because the virtual machines all run in similar virtualization farms.

SLA metrics for the business have impact on requirements on different layers, e.g.:

- Application Layer, e.g. business critical applications preferably using reusable application clusters (e.g. Web, App, and DB). Reusable for other applications and therefore preventing the need for setting up a complete HA/DR environment per application.
- Middleware Layer, e.g. databases, AD, critical for business critical applications. Also must be active/active on site 1 and active (preferred) or passive on site 2.
- IaaS Layer, e.g. virtual machines, can have the lowest provider SLA when the application and middleware already are high available by design.

Requirements to achieve High Availability (HA) by design within one datacenter:

- Silver & Bronze (non-business critical applications): One Intel Virtual Machine is good enough. The machine will restart automatically when the hardware fails.
- Gold: Intel requires at least two (clustered) machines on one site.
- Gold: One IBM Power machine on one site can be seen as HA.

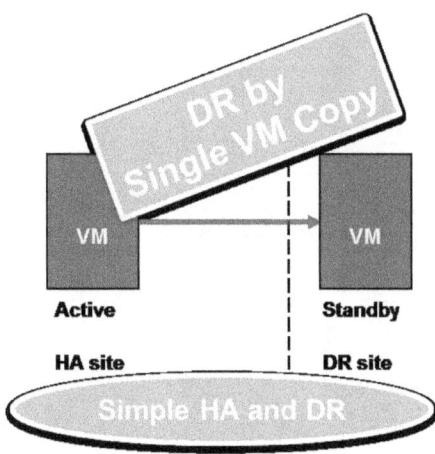 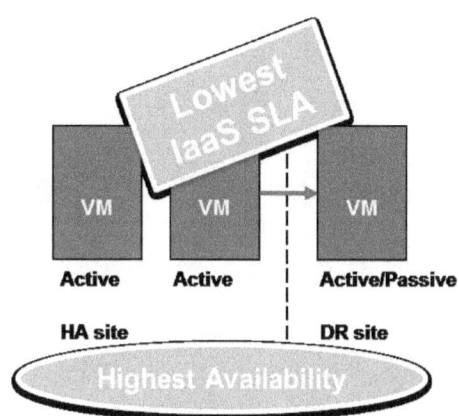

**Lowest SLA (lowest costs) for highest availability**

The left image shows the normal virtual machine setup for providing HA and DR within a standard server virtualization farm and good for an IBM Power environment (AIX, i) for offering availability, continuity, vertical and horizontal scalability. These "Single VM" non-functional features are more limited in an Intel environment.

The right image shows a potential setup for delivering the highest "Intel" SLA for the business with the lowest SLA from the provider. This design offers availability, continuity, horizontal and limited vertical scalability.

Example implementations to achieve Availability and Continuity metrics for the business using the lowest SLA metrics for IaaS:

| SLA | "Gold" | "Silver" | "Bronze" |
|---|---|---|---|
| **HA** **(within 1 DC)** | **For business critical appl. & middleware:**<br>• Active/Active or<br>• Active/Passive<br>• Multi Master RW's<br><br>**Lowest IaaS SLA** | Single VM (auto restart on other physical machine)<br><br><br><br>Medium IaaS SLA | Single VM (auto restart on other physical machine)<br><br><br><br>Lowest IaaS SLA |
| **DR** **(site failover)** | Active or Passive on the other site<br><br>**Lowest IaaS SLA** | Copy (Continuous replication to other DC + e.g. SRM)<br>Medium IaaS SLA | Copy (Continuous replication + e.g. SRM)<br>Lowest IaaS SLA |

### 3.3.4    C3. MS - Description of services

| Details of your products/services | | |
|------|--------------------------------|--------|
| **ID** | **Question** | **Answer** |
| 1. | What are your core activities and what percentage of your sales do they represent? | |
| 2. | In which year did you begin operating in the market with the offered solutions? | |
| 3. | How many similar contracts have you already carried out in Belgium? | |
| 4. | Do you carry out contracts under your own management or in co-operation with other partners? | |
| 5. | Do you outsource essential parts? If so, which? | |
| 6. | When do you involve partners? | |
| 7. | Can you guarantee unreservedly that your company can supply, install and implement all the required functionality (turnkey delivery)? So that there is a single point of contact. Can you also give the same guarantee for maintenance and management? | |
| 8. | Can you also give this guarantee if in time Customer decides to expand the initial supply? | |
| 9. | How can you help us to rapidly onboard new Trading partners? | |

| Quality Management | | |
|------|--------------------------------|--------|
| **ID** | **Question** | **Answer** |
| 10. | Is your organization ISO-certified or do you have a quality assurance system? | |
| 11. | For how long have you held a quality certificate? | |
| 12. | Can you present a report of a quality audit by an external expert confirming that you have a quality system which is at least 70% ISO-compliant? If so, state the ISO standard | |
| 13. | If you hold an ISO certificate, send a copy of it. | |

| Quality Management | | |
| --- | --- | --- |
| **ID** | **Question** | **Answer** |
| **14.** | If you cannot present a report of a quality audit by an external expert, are you prepared to meet the costs of a quality audit carried out by or on behalf of Customer? | |
| **15.** | Do you periodically record and analyze projects which have been delivered too late and what as a rule are the corrective measures (ad hoc and structural)? | |
| **16.** | Can you give details of your complaints procedure and state which projects (past two years) have been involved and how they have been dealt with? | |

| Audit and Information Security | | |
| --- | --- | --- |
| **ID** | **Question** | **Answer** |
| **17.** | The vendor shall ensure, Customer for free, in-control certificate issued by an independent party on the correct, complete and timely delivery of contracted services as well as the management and security of these services. | |
| **18.** | Customer is entitled to a free audit / risk audit on premises of supplier. The supplier shall demonstrate that adequate information security measures have been taken and / or at what time they were implemented. | |
| **19.** | Describe what measures the supplier has taken to prevent unwanted access, changes and availability of data provided by the client. | |
| **20.** | Storage of business information: describe how a data set is stored on the infrastructure of the supplier (outside EU only with permission). | |

Proposed roadmap:

| Proposed roadmap to managed services | | |
|---|---|---|
| ID | Question | Answer |
| **What is you recommendation/plan for handover of responsibility. At what point the datacenter services will be handled by Vendor?:** | | |
| 21. | Lot 2, SAP | |
| 22. | Lot 3, VMware | |
| 23. | Lot 4, Citrix | |
| 24. | Lot 5, Mail | |

Training:

| Training for managed services | | |
|---|---|---|
| ID | Question | Answer |
| **Is there any need for training and is it included within the pricing proposal?:** | | |
| 25. | General governance | |
| 26. | Lot 2, SAP | |
| 27. | Lot 3, VMware | |
| 28. | Lot 4, Citrix | |
| 29. | Lot 5, Mail | |

### 3.3.4.1 C3 - Lot 2, SAP – Description of Services

| SAP migration – Highest SLA – Lowest costs | | |
|---|---|---|
| **ID** | **Question** | **Answer** |
| **1.** | Do you plan to migrate the SAP environment to Intel? | |
| **2.** | If yes, do you plan migrate each Power "virtual machine" into a PaaS cluster and can Customer profit from the lowest IaaS costs (see C2) and still achieve the highest SLA?  | |

| Description of SAP Services | |
|---|---|
| **ID** | **Question** |
| **3.** | Describe your services that most closely matches the current situation and how you comply or outperform. |
| **4.** | Deliver an overview of Customer's machines/storage groups with your selected SLA level (refer to your described SLA categories in C2, Q1, e.g. "gold", "bronze"). |
| **5.** | What services, changes, life cycle management actions are included in the defined managed service monthly recurring costs (in template A)? |
| **6.** | What services, changes, life cycle management actions are **NOT** included in the defined managed service monthly recurring costs? Please deliver an overview of costs per change in your complete pricing catalog (to be delivered with your answer) |
| **7.** | Customer uses detailed real time and monthly monitoring metrics. What do you offer Customer for Real time performance monitoring? |

How do you achieve the following requirements? Describe the architecture of HA, DR, and retention solutions.

| SAP Requirements (Must haves, Should haves, Could haves, Won't haves) | | | |
|---|---|---|---|
| **ID** | **Requirement** | **MoSCoW** | **Answer** |
| **8.** | Production Environment | Must | |
| **9.** | DTA (OTA) Environment | Must | |
| **10.** | Regular site failover test & frequency | Must | |

| SAP Requirements (Must haves, Should haves, Could haves, Won't haves) | | | |
|---|---|---|---|
| ID | Requirement | MoSCoW | Answer |
| 11. | Regular restore test & frequency | Must | |

### 3.3.4.2 C3 - Lot 3, VMware – Description of Services

| ID | Question |
|---|---|
| **Description of VMware Services** | |
| 1. | Describe your services that most closely matches the current situation and how you comply or outperform. |
| 2. | Deliver an overview of Customer's machines/storage groups with your selected SLA level (refer to your described SLA categories in C2, Q1, e.g. "gold", "bronze"). |
| 3. | What services, changes, life cycle management actions are included in the defined managed service monthly recurring costs (in template A)? |
| 4. | What services, changes, life cycle management actions are **NOT** included in the defined managed service monthly recurring costs? Please deliver an overview of costs per change in your complete pricing catalog (to be delivered with your answer) |
| 5. | Customer uses detailed real time and monthly monitoring metrics. What do you offer Customer for Real time performance monitoring? |

How do you achieve the following requirements? Describe the architecture of HA, DR, and retention solutions.

| ID | Requirement | MoSCoW | Answer |
|---|---|---|---|
| **VMware Requirements (Must haves, Should haves, Could haves, Won't haves)** | | | |
| 6. | Production Environment | Must | |
| 7. | DTA (OTA) Environment | Must | |
| 8. | Regular site failover test & frequency | Must | |
| 9. | Regular restore test & frequency | Must | |

### 3.3.4.3 C3 - Lot 4, Citrix – Description of Services

| ID | Question |
|---|---|
| **Description of Citrix Services** | |
| 10. | Describe your services that most closely matches the current situation and how you comply or outperform. |
| 11. | Deliver an overview of Customer's machines/storage groups with your selected SLA level (refer to your described SLA categories in C2, Q1, e.g. "gold", "bronze"). |
| 12. | What services, changes, life cycle management actions are included in the defined managed service monthly recurring costs (in template A)? |
| 13. | What services, changes, life cycle management actions are **NOT** included in the defined managed service monthly recurring costs? Please deliver an overview of costs per change in your complete pricing catalog (to be delivered with your answer) |
| 14. | Customer uses detailed real time and monthly monitoring metrics. What do you offer Customer for Real time performance monitoring? |

How do you achieve the following requirements? Describe the architecture of HA, DR, and retention solutions.

| Citrix Requirements (Must haves, Should haves, Could haves, Won't haves) |
|---|

| ID | Requirement | MoSCoW | Answer |
|---|---|---|---|
| 15. | Production Environment | Must | |
| 16. | DTA (OTA) Environment | Must | |
| 17. | Regular site failover test & frequency | Must | |
| 18. | Regular restore test & frequency | Must | |

Do you guarantee the following non-functional requirements, which are historically proven in the current Customer environment, or do you keep to your standard SLA metrics?

| Citrix Requirements (Must haves, Should haves, Could haves, Won't haves) | | | |
|---|---|---|---|
| ID | Requirement | MoSCoW | Answer |
| 19. | Maximum login time <= 40 sec (100%) | Must | |
| 20. | Keep disconnected session available for 6 hours (100%) | Must | |
| 21. | Reconnect to disconnected session <= 10 seconds (100%) | Must | |
| 22. | Time to grant ore remove user access to virtual desktop or application <= 8 hours (100%) | Must | |
| 23. | Application time to launch <= 30 seconds | Must | |
| 24. | Time to publish a new or updated application after testing and Customer approval <= 8 hours | Must | |

### 3.3.4.4  C3 - Lot 5, Mail – Description of Services

| ID | Question |
|----|----------|
| **Description of Mail Services** | |
| **1.** | Describe your services that most closely matches the current situation and how you comply or outperform. |
| **2.** | Deliver an overview of Customer's machines/storage groups with your selected SLA level (refer to your described SLA categories in C2, Q1, e.g. "gold", "bronze"). |
| **3.** | What services, changes, life cycle management actions are included in the defined managed service monthly recurring costs (in template A)? |
| **4.** | What services, changes, life cycle management actions are **NOT** included in the defined managed service monthly recurring costs? Please deliver an overview of costs per change in your complete pricing catalog (to be delivered with your answer) |
| **5.** | Customer uses detailed real time and monthly monitoring metrics. What do you offer Customer for Real time performance monitoring? |

How do you achieve the following requirements? Describe the architecture of HA, DR, and retention solutions.

| ID | Requirement | MoSCoW | Answer |
|----|-------------|--------|--------|
| **Mail Requirements (Must haves, Should haves, Could haves, Won't haves)** | | | |
| **6.** | Production Environment | Must | |
| **7.** | DTA (OTA) Environment | Must | |
| **8.** | Regular site failover test & frequency | Must | |
| **9.** | Regular restore test & frequency | Must | |

### 3.3.5    C4. MS - Migration Project Organization

| Turnkey delivery | | |
| --- | --- | --- |
| ID | Question | Answer |
| 1. | Describe your portfolio | |
| 2. | How many people do you have available for implementation and installation work and what percentage of these people have already been allocated to other projects from June to Nov 20xx? | |
| 3. | How many people do you have available for maintenance and management activities and what percentage of these people have already been allocated to other projects in 20xx? | |
| 4. | State what the added value of your company is compared to your competitors in this sector. | |
| 5. | (V2) Describe (or add details in an appendix) your action plan for the turnkey delivery, including a schedule and all deliverables, and the subsequent maintenance and management phase. We also expect a proposal for a test plan and a full description of the documentation to be supplied. | To be delivered in V2. |
| 6. | (V2) On the basis of the enclosed service level agreement, give a full description of the work to be carried out and confirmation of the requested service levels. | To be delivered in V2. |

### 3.3.6    C5. MS - Project Methodology, Tools, Technology

Describe your Project Methodology, Tools and Technology used to execute the migration projects.

### 3.3.7    C6. MS - Demonstrated Capabilities and References (template B) (V1)

Embed the template within this paragraph of your answer.

### 3.3.8    C7. MS - Risk File (template C2)

Embed the template within this paragraph of your answer.

### 3.3.9 C8. MS - Capabilities and competences per technical domain

Capability to prevent the need for site failover:

| Location Requirements (Must haves, Should haves, Could haves, Won't haves) | | | |
|---|---|---|---|
| ID | Requirement | MoSCoW | Answer |
| 1. | Are your twin datacenters close enough to each other (< 40km) and with enough throughput capacity to facilitate "active-active" spread over 2 datacenters (see example diagram in C2) and what is the potential impact on price if active/active over 2 sites is used? | Must | |

To overcome (Citrix or SAP) latency issues for Customer's worldwide users:

| Location Requirements (Must haves, Should haves, Could haves, Won't haves) | | | |
|---|---|---|---|
| ID | Requirement | MoSCoW | Answer |
| 2. | What is your datacenter presence in Europe? | Must | |
| 3. | What is your datacenter presence in North America? | Could | |
| 4. | What is your datacenter presence in South America? | Could | |
| 5. | What is your datacenter presence in Asia / Pacific? | Could | |
| 6. | What is your datacenter presence in Australia? | Could | |
| 7. | What is your datacenter presence in Africa? | Could | |
| 8. | Citrix requires a maximum latency of 200 msec. How do you plan to comply? | Must | |

Virtual datacenter capability.
Extending vendor's datacenters with redundant connectivity to (private) cloud services:

| Location Requirements (Must haves, Should haves, Could haves, Won't haves) | | | |
|---|---|---|---|
| ID | Requirement | MoSCoW | Answer |
| 9. | Do you offer a transparent and high available connection to MS Azure with QoS? | Should | |

| Location Requirements (Must haves, Should haves, Could haves, Won't haves) | | | |
|---|---|---|---|
| **ID** | **Requirement** | **MoSCoW** | **Answer** |
| 10. | Do you offer a transparent and high available connection to Amazon WS with QoS? | Should | |

Related costs are requested in the pricing template A, excluding bandwidth costs with the assumption we can extend our Internet bandwidth using the given bandwidth pricing.

What is the defined skillset of your FTE expert pool available for Customer and what is the number of FTE's available for assignment to Customer, per skill area?:

| Available FTE's for managed services | | |
|---|---|---|
| **ID** | **Question** | **Answer** |
| 11. | General governance | |
| 12. | Lot 2, SAP | |
| 13. | Lot 3, VMware | |
| 14. | Lot 4, Citrix | |
| 15. | Lot 5, Mail | |

Dependency overview to analyze impact of changes:

| Risk mitigation for change management | | |
|---|---|---|
| **ID** | **Question** | **Answer** |
| 16. | What tooling do you use to get an overview of the dependency chain of applications and servers? | |

### 3.3.10   C9. MS - Opportunities Form (template E)

Embed the template within this paragraph of your answer. The opportunities will be used to determine added value of the vendor.

### 3.3.11   C10. MS - RACI's, policies, procedures (V2)

Expected delivery during V2.
The RACI's will be used to clarify responsibilities and tasks for the retained team at Customer.

### 3.3.12   C11. MS - Project Plan & Roadmap (V3)

V1&V2: High level estimation of total project turnaround period per lot and costs in part A
V3: Project plan proposal and estimated man days per task and total project turnaround period per lot.

### 3.3.13   C12. MS - Interviews and PoC / demo (between V3-V4)

Expected delivery during V3.
Between vendor activity periods V3 & V4, Customer requests a live demo, and interviews (to be recorded) with key engineers, planned for Customer, involved in the day-to-day managed services.

For interview without sales presence: Select 2 managed service engineers, who will actually be assigned to Customer. What are the names, and role descriptions of the proposed interview candidates?

### 3.4 PART D. HR TRACK

Expected delivery during V1.

#### 3.4.1    D1. HR - CLA 32bis Experience and References (template D)

Expected delivery during V1:
Show capability and deliver references applying the Belgian Collective Labor Agreement 32bis. Embed template D for your answer.

Also answer the following high level indicators about HR handover experience within the last years and your opinion of potential reduction of Customer employees:

| HR handover experience (2012-...) | | |
|---|---|---|
| ID | Question | Answer |
| 1. | How many Belgian customers did a handover of FTE's? | |
| 2. | How many FTE's were involved? | |
| 3. | What are the potential vendor locations where the Customer engineers could be transferred to? | |
| 4. | How many are still working for you? | |

Number of potential obsolete FTE's:

| HR handover | | |
|---|---|---|
| ID | Question | Answer |
| 5. | How many Customer FTE's do you propose for handover per technical domain (per Lot)? | |

Expected delivery during V3, based on pending delivery of HR details before V3:

*   HR track, per person, e.g.
    *   Pension
    *   Monetary Value calculation from old to new situation

Vendor may deliver a checklist during V1 with required information needed to be able to create a detailed personnel offer in V3.

## 4   TERMINOLOGY / ABBREVIATIONS / GLOSSARY

| Term/Abbreviation | Description |
|---|---|
| AD | Active Directory (for Identity and Access Management (IAM), constraint for almost all services) |
| App | Application |
| Availability Zone | Isolated location for a datacenter, including power supply and cooling. The other Availability Zone is expected 10-40 km apart. |
| CLA | Collective Labor Agreement (Dutch: CAO) |
| DB | Database |
| DC | Datacenter |
| DTA | Development, Test, Acceptance |
| ERP | Enterprise Resource Planning |
| ESB | Enterprise Service Bus for Application Integration (requires a managed meta directory). (Not within the RFP scope) |
| HR | Human Resources |
| IaaS | Infrastructure-as-a-Service, e.g. a managed virtual machine including operating system |
| IAM | Identity and Access Management |
| Cus | Location 1 |
| KPI | Key Performance Indicators |
| L&S, LS | Lift & Shift |
| MRC | Monthly Recurring Cost |
| MS | Managed Service |
| NDA | Non-Disclosure Agreement |
| Ops | Operations |
| OS | Operating System |
| OTA | Development (Ontwikkel), Test, Acceptance |
| OTC | One Time Cost |
| P1-P4 | Severity levels within SLA, Priority 1 – 4, e.g. P1 is major incident. |
| PaaS | Platform-as-a-Service, e.g. IaaS environment including managed middleware such as database or application development environment. |
| PoC | Proof of Concept |
| QoS | Quality of Service (to guarantee defined bandwidth) |
| RACI | Roles & Responsibilities matrix: Responsible, Accountable, to be Consulted, to be Informed |
| RFI | Request for Information |
| RFP | Request for Proposal |
| RFQ | Request for Quote |
| RPO | Recovery Point Objective (How much data is allowed to get lost in minutes/hours/days) |
| RTO | Recovery Time Objective (Resolution time: When are services up again in maximum minutes/hours/days) |
| RW's | Multi master Read Write Replica's, such as used in a HA AD design. |
| SaaS | Software-as-a-Service, e.g. a managed application on top of PaaS. |
| SLA | Service Level Agreement |
| SOC | Secure Operating Center (kind of Intrusion Detection System) |
| SPOC | Single Point Of Contact |
| V1-V4 | Vendor activity periods to enable vendor resource planning for RFP answering |

| VM | Virtual Machine |
| WAF | Web Application Firewall |

## 5   APPENDIX A: DOCUMENTS OVERVIEW

Document list for the complete RFP:

- NDA (pre RFP document)
- This document: Datacenter Hosting RFP
- Template A – Pricing (to keep external)
- Template B – Demonstrated Capabilities and References
- Template C1 – Risk Reference File – Lift & Shift
- Template C2 – Risk Reference File – Managed Services
- Template D – HR Experience and References
- Template E – Opportunities
- Information Library (Due Diligence Inventory docs/folders):
  - Lot 1, datacenter
    - IP Plan
    - Network design diagrams (DC, WAN)
  - Lot 2, SAP
  - Lot 3, VMware
  - Lot 4, Citrix
  - Lot 5, Mail
  - General:
    - Operator handbooks
    - Process and Procedures Manuals
    - Service Level Reporting of the last months
    - Third party contracts
    - Main licenses
    - Monthly reports – al domain reports during an example month.

### 5.1 Download – Escpecially for readers of this book

You may download the following templates:

- Demonstrated Capabilities and References Template
- Opportunities Template
- Pricing Template
- Risk and Reference Template
- Surprise Template making visits of IT consultants and engineers more effective

http://www.cioforum.nl/cioforum/english/bvp/

## 6   DOCUMENT INFORMATION

### 6.1 Revision history

| Date | Name | Summary of Changes | Version |
|------|------|--------------------|---------|
| <date> | Robert Zondervan | Final version for RFP V1 | v1-00 |
| | | | |
| | | | |
| | | | |
| | | | |

## 6.2 About the Author

I like to help IT directors and CIOs who want to be sure that their interest comes first in working with IT suppliers.

I believe I can contribute to your success. To help you in choosing the best solutions based on verifiable requirements.

After more than 15 years experience in explaining IT infrastructure solutions to engineers, I now have more than 15 years experience in choosing and explaining solutions to business owners and governments.

I will take you into an experience where you are in control and where you only select the most valuable IT investments, taking into account the main problems, needs and risks with the least amount of resources.

**Robert Zondervan, MSc**
*Founder and*
*Shared IT-director of*
*CIOforum.nl*

*Master in ICT Enterprise*
*Architecture*

*Certified in the Governance*
*of Enterprise IT,*
*Isaca cert nr 1406183*

## 6.3 Other publications from CIOforum.nl

Govern Your IT
ISBN-13: 978-1534603448
ISBN-10: 1534603441
Publication date: September 2016

**Govern your IT**
Would it not be great if your IT did exactly what your organization wants and needs it to do? Surely, you want to invest only in projects that add the most value, tackle your biggest risks, use the most efficient building blocks and processes, and that cost you the least. This is where my AMORT system comes in, which will have you achieve success in a matter of days. In this book, I will use plain language to present short practical steps and surprising tips and ideas that will make life a lot easier for managers, owners, and directors. The ideal is a situation where your IT is flexible and scalable, keeping your costs down during both busy and quiet periods.

Where ad-hoc purchases and short-term investments are keenly aligned with your long-term vision. Where you have a point on the horizon to gradually work towards. Where your business strategy has been translated into the ideal strategy for your IT. To achieve this ideal, read this book. And after reading this book, you can fully focus on your organization, on your core expertise, because your IT does exactly what you want and need it to do.